Scattering and Diffraction by Wedges 1

To my beloved grandchildren Marianna and Roberto
V.G.D.

To my family Stefania, Marina and Alessandro for their patience and support, this book is affectionately dedicated
G.L.

Waves and Scattering Set

coordinated by
Jean-Michel L. Bernard

Volume 1

Scattering and Diffraction by Wedges 1

The Wiener-Hopf Solution – Theory

Vito G. Daniele
Guido Lombardi

iSTE

WILEY

First published 2020 in Great Britain and the United States by ISTE Ltd and John Wiley & Sons, Inc.

ISTE Ltd
27-37 St George's Road
London SW19 4EU
UK

www.iste.co.uk

John Wiley & Sons, Inc.
111 River Street
Hoboken, NJ 07030
USA

www.wiley.com

Library of Congress Control Number: 2020938054

British Library Cataloguing-in-Publication Data
A CIP record for this book is available from the British Library
ISBN 978-1-78630-263-2

Contents

Preface

The theory of wave diffraction by wedges constitutes one of the fundamental problems in mathematical physics. Beyond the obvious applications to engineering and physics, this topic has generated – and continues to generate – new progress in the techniques of applied mathematics (e.g. the Sommerfeld–Malyuzhinets technique and the Kontorovich–Lebedev method), and applied physics (e.g. the geometrical theory of diffraction and the physical theory of diffraction).

The scattering of wedge structures is of interest in the study of electromagnetic response/interaction of more complex, multiscale computational electromagnetic problems, using diffraction coefficients for modeling subdomains where high-frequency methods hold promise.

This book, which is divided into two volumes, presents a general flexible, novel and powerful methodology for solving problems of electromagnetism in regions containing arbitrary wedge objects; this technique can be extended, as suggested, to different physics.

The proposed technique is based on a generalization of the Wiener-Hopf (WH) technique that allows the study of complex canonical scattering problems.

Research carried out during the past two decades has ordered and systematized the procedure, to obtain spectral equations and integral representations for complex problems to avoid redundancy.

The purpose of this book is to present the application of the so-called generalized Wiener-Hopf technique (GWHT) to wedge scattering problems. This method often adopts very sophisticated mathematical methods; however, since the aim of the authors is to favor the acquisition of fundamental concepts in view of possible applications, they prefer to omit rigorous mathematical proofs for theory such as theorems. For all unproven mathematics, we refer to papers and books that are easily available.

Due to the electrical engineering background of the authors, the two volumes are focused on electromagnetic applications. However all of the material is presented in such a way that extension to other physics and applied mathematical fields is straightforward; from applications in all areas of diffraction problems, such as acoustics, elasticity, aerodynamics, hydrodynamics and so on.

Analogies between some resolved problems in electromagnetics and other areas of physics are reported. In particular, for example, the diffraction of a half-infinite crack in an elastic solid medium is described in detail.

Before dealing with the applications to wedge problems, we present some useful remarks about the WH technique. First, the classical and consolidated method of solution of WH equations is founded on a deep knowledge of the properties of complex functions, which is based on the following classical steps: the multiplicative factorization of the kernel, the additive factorization (decomposition) of functions and the application of Liouville's theorem.

Second, in general, no WH problem is simple to study and solve. The formulation can be cumbersome to obtain and very specialized techniques are needed to implement the factorization procedure. Moreover, these methods frequently do not allow small variations of the problem itself. In this context, it is fundamental to adopt spectral domains using spatial Laplace transforms where all the unknown field representations are analytic functions.

Furthermore, even though formal solutions can be obtained in particular cases, a long and difficult elaboration may be required to make them effective from the physical and engineering points of view. For this purpose, tools such as the saddle point method are necessary for the asymptotic evaluation of integrals.

In this book, we aim to solve these problems systematically by helping the reader to implement solutions using the WH method with precise, simple steps and reducing the restrictions on generalization.

In particular, we present one of the most relevant recent advances in the WH technique, i.e. the transformation of WH equations into integral equations for their effective solution via simple discretization.

This method, known as the Fredholm factorization, is an approximate technique that extends the applicability of the WH method to a new variety of problems, consisting of different geometries and materials.

The Fredholm factorization is a semi-analytical solution method that preserves the spectral properties of the problem through its structural singularities and singularities related to the sources. Approximations are present on correction terms that are only related to diffractive properties. Thus, using a suitable analytic continuation, the method allows the reader to look at the physical properties of the problems, as done in analytical closed-form solution methods.

The success of the Fredholm factorization, in solving complex diffraction problems involving isotropic media, has also encouraged the authors to extend the WH technique developed in this book to wedge problems involving anisotropic or, more general, bi-anisotropic media. Moreover, in these cases, the Fredholm factorization is conceptually possible. However, in most cases, it yields complex kernels that involve cumbersome multivalued functions. Much effort is needed to complete the general WH theory for solving wedge problems involving arbitrary linear media. In particular, difficulties increase considerably when it is not possible to take advantage of the presence of Helmholtz equations. In fact, in these cases, it is very difficult and cumbersome to implement the indispensable analytic continuation procedures.

However, to successfully develop the mathematical theory and applications of this book, it is useful to have a technical computing software, particularly in analytical and numerical manipulations of the equations.

Finally, we admit that the authors are strongly influenced by the fundamental work of Felsen and Marcuvitz on radiation and scattering by waves, and therefore the adopted notations follow their studies. In particular, equivalent network representations of equations are provided to systemize the mathematical procedure. The use of different notations in applied physics, applied mathematics and engineering communities should not present a major difficulty in reading this book.

Vito G. DANIELE
Guido LOMBARDI
May 2020

Introduction

Diffraction by objects presenting edges is a fundamental problem in the theory of diffraction. Its importance has been demonstrated by the vast number of papers and books that deal with this topic. These works concern mathematical and physical aspects, with applications to many branches of physics and engineering such as electromagnetism, acoustics, elasticity, aerodynamics and hydrodynamics. To provide a complete theory of the effects of the edges present in a body, we study wedge problems, i.e. canonical problems where the diffracting body is composed of homogeneous wedges. Wedge problems can be classified in different ways. For example, in electromagnetism, we introduce impenetrable and penetrable wedges. Impenetrable wedges arise from the introduction of approximate boundary conditions on the surfaces of wedges. For example, near a metallic face, at microwave frequencies, we can assume that the tangential electrical field vanishes. The introduction of impenetrable wedges simplifies the study of wedge problems considerably since the external problem (i.e. the evaluation of the electromagnetic field outside the wedge) is decoupled from the internal problem (i.e. the evaluation of the electromagnetic field inside the wedge). Considerable advances have been made to obtain more realistic approximate boundary conditions. Unfortunately, the introduction of sophisticated boundary conditions is not sufficient to study penetrable wedges. In particular, the scattering by a penetrable wedge (e.g. the dielectric wedge in electromagnetics) is a very difficult problem that has arisen for a long time and drawn the interest of many scientists.

Nevertheless, even for the simplest cases, no closed-form solution for this problem is known.

The solutions of wedge problems have a long history, which have been well documented by numerous books and papers. This history also documents the importance of the use of spectral representations to obtain the solutions of wedge problems. One of the aims of these two volumes is to claim the power of the representations of the field spectra through Laplace transforms in radial direction. In particular, the use of these representations allows the extension of the powerful WH technique to formulate and efficiently solve many wedge problems.

This book is divided into two volumes that are consistent with each other. The first volume introduces the general theory with simple examples, while the second volume reports more advanced applications. Each volume consists of three chapters and their cross-references.

In each volume, sections, equations and figures are numbered as (chapter.section.item) in order for the content to be easily accessible.

Chapter 1 starts by recalling some fundamental properties of the mathematical theory of the WH technique. The separation of the plus and minus Laplace transforms is fundamental in introducing the concept of the classical factorization, an ingenious idea forming the basis of this technique. This chapter presents a few cases where exact factorizations are available. A novel aspect presented in this chapter is the introduction of the solution procedure for the Fredholm factorization, as an alternative to the classical WH solution procedure. It presents the fundamental advantage of its applicability in general cases where the classical factorizations are not available.

To discuss the importance of the mathematical tools described in Chapter 1, the solution of the most simple diffraction problem is presented in Chapter 2: the electromagnetic scattering of a plane wave by a perfectly electrically conducting (PEC) half-plane. Both the classical factorization procedure and the Fredholm factorization method are applied to obtain the solution of this problem. The comparison of these two different factorization techniques illustrates their advantages and disadvantages. To provide a general methodology for studying basic problems in other diffraction areas, Chapter 2 presents the complete theory of the diffraction of a half-infinite crack located in an indefinite elastic solid.

Chapter 3 deals with the application of the WH technique to wedge problems. This application comes from the introduction of generalized Wiener-Hopf equations (GWHEs) and from the mapping that reduces GWHEs to classical WH equations (CWHEs).

In order to compare the WH technique with the Sommerfeld–Malyuzhinets technique, this chapter introduces two angular complex planes called w and bar w. In this framework, we present a new tool called rotating waves to study angular regions. Moreover, the introduction of angular complex planes is also useful in developing the application of the Fredholm factorization method.

Chapter 4 (Volume 2) deals with the solution of impenetrable wedges immersed in free space. Exact solutions obtained with the classical factorization theory are compared with those obtained by the Sommerfeld–Malyuzhinets method.

Chapter 5 (Volume 2) is particularly important for the development of immediate engineering applications. In fact, it presents the application of the Fredholm factorization method to the solution of the general scattering problem by an arbitrary impedance wedge illuminated by plane waves at skew incidence.

Finally, Chapter 6 (Volume 2) deals with the use of the WH technique to solve an arbitrary homogeneous penetrable wedge immersed in free space. In particular, this chapter studies, in depth, the diffraction by a dielectric wedge illuminated by plane waves at normal and skew incidence. The WH solution of the dielectric wedge problem presented in this chapter is the first one that the authors of this book have developed in the literature. It contains very sophisticated mathematical methods to overcome difficult analytical problems related to the presence of two media. Recently, many improvements have been made to the WH method to study the diffraction of complex problems consisting of multiple wedges and layers (see 2016–2019 papers by the authors of this book). We expect that these improvements will allow a more simple mathematical procedure to solve the diffraction by a dielectric wedge.

1

Introduction to the Wiener-Hopf Method

1.1. A brief history of the Wiener-Hopf method

In 1931, Wiener and Hopf (Wiener and Hopf 1931) invented a powerful technique for solving a special type of integral equation. By introducing the Laplace transform of the unknown, the integral equation was rephrased in terms of a functional equation in a suitably defined complex space. The solution method was ingenious and based on a sophisticated procedure that exploits some properties of analytic functions. It stands as one of the most important mathematical inventions for obtaining analytical solutions of very difficult problems. The number of problems that can be effectively approached by the Wiener-Hopf (WH) technique is significant. Most of these problems are very important, and often there are no alternative approaches to solve them efficiently.

In this chapter, we recall the notations and the fundamental concepts of this technique. For a comprehensive exposition of the theory, the reader can also refer to Noble (1958) and Daniele and Zich (2014). Moreover, we introduce a new methodology for the semi-analytical solution of classical WH equations, i.e. the Fredholm factorization method, which will be used in generalized form in the following chapters.

1.2. Fundamental definitions and assumptions to develop the Wiener-Hopf technique in the spectral domain

Throughout this book, we will only consider time harmonic fields with a time dependence specified by the factor $e^{j\omega t}$ (electrical engineering notations), which is

---

This book consists of two volumes; any references to Chapters 4, 5 or 6 can be found in the second volume: *Scattering and Diffraction by Wedges 2*.

omitted, where the imaginary unit is indicated by j. Applied mathematics often uses a different notation $e^{-i\omega t}$, where the imaginary unit is indicated by i. This means that, in the natural domain, the change $j \Rightarrow -i$ transforms engineering notations in applied mathematics notations (and vice versa).

We recall that by the use of the Laplace (or Fourier transform) we define the diffraction problem in a spectral domain. In the spectral domain we need to pay particular attention to the notations used in engineering and applied mathematics (see Appendix 1.A for a discussion on the use of Fourier transforms extendable to Laplace transforms).

Due to mathematical reasons related to the WH technique, the ideal media used for describing the problems will be considered with small vanishing loss. Taking into account the engineering notation, it means that their propagation constants (for example) k has a negative (vanishing) imaginary part (see Appendix 1.A).

Throughout this book, only remote sources consisting of plane waves will be considered. As is well known, this limitation is not restrictive because, due to the linearity of the considered problems, a general source can always be expanded in terms of plane waves (Felsen and Marcuvitz 1973). In this book, a fundamental role is played by the geometrical optical (GO) contributions in the presence of incident plane waves. As these contributions are well known from elementary considerations, we assume that the reader is able to understand them.

The separation of the plus and minus Laplace transforms is fundamental to formulate the problem in terms of spectral WH functional equations, which forms the basis of the concept of the classical factorization in the WH technique.

We introduce the spectral domain η with the following integral definitions of plus and minus WH functions that are unknown at the beginning of the procedure:

$$F_+(\eta) = \int_0^\infty e^{j\eta x} f(x)dx \text{ or } F_-(\eta) = \int_{-\infty}^0 e^{j\eta x} f(x)dx \tag{1.2.1}$$

The above integrals converge for values of η located in an upper half-plane (the first integral) or a lower half-plane (the second integral). Their analytic continuations for values arbitrary of η define the plus and minus Laplace transforms. The singularities of the plus or minus function can be poles or branch points. For example, in the spectral domain, the GO contributions only produce poles.

While, in the presence of sources located in finite points, the plus (minus) functions are regular in the upper (lower) half-plane $\text{Im}[\eta] \geq 0$ ($\text{Im}[\eta] \leq 0$), the presence of plane waves and the losses assumed in the medium can produce non-standard poles. For a plus (minus) function, we define non-standard poles located in $\mathrm{Im}[\eta]\geq 0$ ($\mathrm{Im}[\eta]\leq 0$). In general, we can decompose the plus and minus functions in standard and non-standard parts:

$$F_{+,-}(\eta)=F_{+,-}^{s}(\eta)+F_{+,-}^{n.s}(\eta) \tag{1.2.2}$$

where:

– $F_{+}^{s}(\eta)$ is the standard part of a plus function that is regular in the upper half-plane $Im[\eta]\geq 0$;

– $F_{-}^{s.}(\eta)$ is the standard part of a minus function that is regular in the lower half-plane $Im[\eta]\leq 0$;

– $F_{+}^{n.s.}(\eta)$ is the non-standard part of plus functions, which consists of the characteristic part of the poles located in the upper half-plane $Im[\eta]>0$;

– $F_{-}^{n.s.}(\eta)$ is the non-standard part of minus functions, which consists of the characteristic part of the poles located in the lower half-plane $Im[\eta]<0$.

If we consider the above Laplace transformations related to total fields, we observe that the non-standard parts of the WH unknown functions are coincident with the non-standard part of the GO contributions. Consequently, non-standard parts are known *a priori*:

$$F_{+,-}^{n.s}(\eta)=\left[F_{+,-}^{G.O}(\eta)\right]^{n.s.} \tag{1.2.3}$$

In this book, wide use of multivalued analytic functions is demonstrated; thus, the determination of branches on Riemann surfaces is a fundamental mathematical issue.

To successfully develop the mathematical theory and applications of this book, it is useful to have technical computing software, particularly in analytic and numerical manipulations of the equations.

In particular, in this book, for wedge scattering problems, multivalued functions are defined in terms of the spectral propagation multivalued function:

$$\xi(\eta)=\sqrt{k^2-\eta^2} \tag{1.2.4}$$

where the proper (principal) sheet is defined for $\xi(0)=k$, thus $\text{Im}[\xi(\eta)]\leq 0$ and the common selection of branches originated from the branch point $\pm k$ are the line with $\text{Im}[\xi]=0$, the vertical lines or the lines with $\text{Re}[\xi]=0$. A discussion on the proper and improper sheets of $\xi(\eta)=\sqrt{k^2-\eta^2}$ is presented in section 1.13.

1.3. WH equations from the physical domain to the spectral domain

WH equations arise from discontinuity problems; therefore, they take the forms of semi-convolutional equations in the physical domain:

$$\int_0^\infty g(x-x')f(x')dx'=f_o(x),\ 0<x<\infty \tag{1.3.1}$$

where f(x) is the unknown, g(x) is the kernel and $f_o(x)$ is a forcing (source) term.

WH equations can be considered to be generalizations of convolutional equations:

$$\int_{-\infty}^\infty g(x-x')f(x')dx'=f_o(x),\quad -\infty<x<\infty \tag{1.3.2}$$

where solutions can be obtained using Fourier transforms in the η plane:

$$F(\eta)=\mathbb{F}[f(x),x,\eta]=\int_{-\infty}^\infty f(x)e^{j\eta x}dx \tag{1.3.3}$$

$$F(\eta)=[G(\eta)]^{-1}F_o(\eta) \tag{1.3.4}$$

The semi-convolutional equation (1.3.1) can be extended in the domain to all real values of x as

$$\int_{-\infty}^\infty g(x-x')f(x')u(x')dx'=f_o(x)u(x)+f_-^s(x)u(-x),\quad -\infty<x<\infty \tag{1.3.5}$$

where an unknown function $f_-^s(x)$ and a plus unknown function $f(x)u(x)$ are introduced, and $u(x)$ is the unit step Heaviside function. The application of Fourier transform to (1.3.5) yields

$$G(\eta)F_+(\eta)=F_-^s(\eta)+F_{o+}(\eta) \tag{1.3.6}$$

where

$$F_+(\eta) = \int_{-\infty}^{\infty} f(z)u(z)e^{j\eta z}dz = L[f(z), z, -j\eta] = \int_0^{\infty} f(z)e^{-sz}dz\Big|_{s=-j\eta} = \int_0^{\infty} f(z)e^{j\eta z}dz$$

$$F_-^s(\eta) = \int_{-\infty}^{\infty} f_-^s(z)u(-z)e^{j\eta z}dz = \int_{-\infty}^{0} f_-^s(z)e^{j\eta z}dz$$

$$F_{o+}(\eta) = \int_{-\infty}^{\infty} f_o(z)u(z)e^{j\eta z}dz$$

(1.3.7)

Equation (1.3.6) is called the non-homogeneous WH equation. The homogeneous WH equation

$$G(\eta)F_+(\eta) = F_-(\eta) \tag{1.3.8}$$

can be defined by absorbing $F_{o+}(\alpha)$ in the WH unknown $F_-(\alpha)$:

$$F_-(\eta) = F_-^s(\eta) + F_{o+}(\eta) \tag{1.3.9}$$

We also note that the physical support of plus unknown is x>0 and that of the minus unknown is x<0.

The standard properties of the plus and minus functions are related to the location of source poles.

1.4. WH equations in wave scattering problems

The WH equations can be classified into scalar or vector. The more general WH equations present the vector form:

$$\mathbf{G}(\eta)\mathbf{F}_+(\eta) = \mathbf{F}_-(\eta) + \mathbf{F}_o(\eta) \tag{1.4.1}$$

where $\mathbf{F}_+(\eta)$ and $\mathbf{F}_-(\eta)$ are respectively plus and minus vector unknowns of order n, $\mathbf{F}_o(\eta)$ is a vector source term of order n and $\mathbf{G}(\eta)$ is the matrix WH kernel of the same order.

Now we introduce a smart property which leads us to select Laplace transforms in the deduction of WH equations.

While considering remote sources such as plane waves in layer-scattering problems, $\mathbf{F}_o(\eta)$ in (1.4.1) is constituted of plane wave contributions that results to be zero in the spectral plane defined by the double-sided Laplace transforms, although its inversion rebuilds the plane waves, Daniele and Zich (2014, Chapter 7.3). On the contrary, using Fourier transforms, plane waves are represented by distributions as $e^{-j\eta_o x} \rightarrow 2\pi\delta(\eta-\eta_o)$ that are cumbersome to be considered throughout the entire development of the WH theory.

For this reason, we deduce the WH equations for layered regions in Laplace domain by ignoring the presence of remote sources. This property can be extended to angular regions (see section 3.3). Consequently, in this framework, the relevant WH equations assume the homogeneous form:

$$\mathbf{G}(\eta)\mathbf{F}_+(\eta) = \mathbf{F}_-(\eta) \tag{1.4.2}$$

This is remarkable since the homogeneous form of the WH equations is more flexible than the general one.

When the WH equations involve multivalued functions in the kernel $\mathbf{G}(\eta)$, we can limit the study of solutions on a Riemann sheet (e.g. the principal sheet). It arises from the fact that even though we define (1.4.2) in a particular sheet and provide a solution in this sheet, this solution consists of an analytic element of the plus and minus functions.

The analytic continuation of these solutions provides the complete multivalued plus and minus functions. Henceforth, all our considerations will refer to the principal sheet of the spectral planes involved in the problems.

The solution of the WH equations is based on the location of the singularities in the principal sheet of the η-plane related to the kernel $\mathbf{G}(\eta)$ and the unknowns $\mathbf{F}_+(\eta)$ and $\mathbf{F}_-(\eta)$.

While considering wave scattering problems, we infer these singularities by considering the contributions of the sources to the problem.

In the case of remote sources represented by plane waves, the GO contributions $\mathbf{F}_+^{GO}(\eta)$ and $\mathbf{F}_-^{GO}(\eta)$ only produce poles (GO poles).

The GO contributions consist of very important approximate solutions of the problems considered in this book, often easily obtained without solving the diffractive effects of the problem.

As stated previously, the non-standard contributions to the Laplace transforms always arise from the GO poles. Consequently, we can evaluate *a priori* the non-standard contributions of $\mathbf{F}_+(\eta)$ and $\mathbf{F}_-(\eta)$ since they are the same as those of $\mathbf{F}_+^{GO}(\eta)$ and $\mathbf{F}_-^{GO}(\eta)$, ($\mathbf{F}_{+,-}^{n.s.}(\eta) = \mathbf{F}_{+,-}^{GOn.s.}(\eta)$).

A special case is when a GO contribution is with $\eta_o = 0$. This pole is not standard for both the plus and minus functions. For example, for this pole, we have:

$$F_+^{n.s.}(\eta) = F_+^{GOn.s}(\eta) = \frac{T_o}{\eta}, \quad F_-^{n.s.}(\eta) = F_-^{GOn.s}(\eta) = \frac{R_o}{\eta} \qquad (1.4.3)$$

where, from (1.4.2), $R_o = G(0)T_o$.

Conversely, the singularities present in the kernel $\mathbf{G}(\eta)$ and its inverse $[\mathbf{G}(\eta)]^{-1}$ do not depend on the source. These singularities (poles and/or branch points) are referred to as structural since they depend only on the geometry of the problem. For example, the free space produces the presence in $\mathbf{G}(\eta)$ and $[\mathbf{G}(\eta)]^{-1}$ of two branch points located in $\eta = \pm k$. Since the media present in the geometries of the problems are always considered with small vanishing loss, no structural singularities are located on the real axis $\mathrm{Im}[\eta] = 0$.

The role of the structural singularities is important. In effect, taking into account (1.4.2), they can contribute to $\mathbf{F}_+(\eta)$ and $\mathbf{F}_-(\eta)$, in addition to the GO poles. In general, these contributions have a physical meaning: waveguide modes, surface waves, leaky waves, lateral waves and so on.

Moreover, the physical interaction between the structure and the sources generate diffraction, whose coefficients can be evaluated by an efficient asymptotic estimation of the integrals that involve the continuous spectrum related to the physical spectrum of the problem.

While formulating the problem, we should note that the field components must satisfy the edge conditions in angular regions near $\rho \to 0$ and the radiation condition at $\rho \to \infty$.

In wedge problems, each field component presents the form $f(\rho) = 0(\rho^\nu)$, where the real value ν depends on the component of the field, the geometry and the materials of the angular region. It is remarkable that in all physical cases, $\nu > -1$ (due to the finiteness of energy). From the Watson lemma, the Laplace transforms

$\mathbf{F}_+(\eta)$ and $\mathbf{F}_-(\eta)$ behave as $O(\eta^{-\nu-1})$ as $\eta \to \infty$, and from (1.4.2) we have $G(\eta)/\eta \to 0$ and $[G(\eta)]^{-1}/\eta \to 0$ as $\eta \to \infty$.

1.5. Classical solution of WH equations

The classical ingenious procedure for the solution of WH equations (1.4.2) is based on three steps: 1) the multiplicative factorization of the kernel, 2) the additive factorization (decomposition) of functions and 3) the application of Liouville's theorem.

The multiplicative factorization of the kernel $\mathbf{G}(\eta)$ can be chosen within a constant factor:

$$G(\eta) = G_-(\eta) G_+(\eta) \tag{1.5.1}$$

The two factorized functions $G_-(\eta)$ and $G_+(\eta)$ along with their inverse $[G_-(\eta)]^{-1}$ and $[G_+(\eta)]^{-1}$ must be regular in $\mathrm{Im}[\eta] \leq 0$ and $\mathrm{Im}[\eta] \geq 0$ respectively. In addition, the factorization (1.5.2) is called regular if $\frac{G_\pm(\eta)}{\eta} \to 0$ and $\frac{[G_\pm(\eta)]^{-1}}{\eta} \to 0$ as $\eta \to \infty$.

From (1.4.2), we get:

$$G_+(\eta) F_+(\eta) = [G_-(\eta)]^{-1} F_-(\eta) \tag{1.5.2}$$

The possible presence of the non-standard part in $\mathbf{F}_+(\eta)$ and $\mathbf{F}_-(\eta)$ prevents the possibility of having the first member regular in $\mathrm{Im}[\eta] \geq 0$ and the second member regular in $\mathrm{Im}[\eta] \leq 0$.

To obtain the regularity of the two complementary members, we introduce the non-standard parts $\mathbf{F}_+^{n.s}(\eta) = \mathbf{F}_+^{GOn.s}(\eta)$ and $\mathbf{F}_-^{n.s}(\eta) = \mathbf{F}_-^{GOn.s.}(\eta)$:

$$\mathbf{F}_+^{n.s}(\eta) = \sum_i \frac{R_{F_+}^{n.s.}(\eta_{ui})}{\eta - \eta_{ui}} + \frac{T_o}{\eta}, \quad \mathbf{F}_-^{n.s}(\eta) = \sum_i \frac{R_{F_-}^{n.s.}(\eta_{li})}{\eta - \eta_{li}} + \frac{R_o}{\eta} \tag{1.5.3}$$

Note that at least one term in (1.5.3) needs to be present due to the presence of sources.

In (1.5.3), η_{ui} and η_{li} are respectively the non-standard poles of $\mathbf{F}_+^{n.s}(\eta) = \mathbf{F}_+^{GOn.s}(\eta)$ with known residues $R_{F_+}^{n.s.}$ and $\mathbf{F}_-^{n.s}(\eta) = \mathbf{F}_-^{GOn.s}(\eta)$ with known residues $R_{F_-}^{n.s.}$.

Let us introduce the functions:

$$C_-(\eta) = \sum_i \frac{G_+(\eta_{ui})R_{F_+}^{n.s.}(\eta_{ui})}{\eta - \eta_{ui}} + \frac{G_+(0)T_o}{2\eta} \tag{1.5.4}$$

$$C_+(\eta) = \sum_i \frac{G_-^{-1}(\eta_{li})R_{F_-}^{n.s.}(\eta_{li})}{\eta - \eta_{li}} + \frac{G_-^{-1}(0)R_o}{2\eta} \tag{1.5.5}$$

and rewrite (1.5.2) in the form

$$G_+(\eta)F_+(\eta) - C(\eta) = [G_-(\eta)]^{-1} F_-(\eta) - C(\eta) = w(\eta) \tag{1.5.6}$$

where:

$$C(\eta) = C_+(\eta) + C_-(\eta) \tag{1.5.7}$$

Taking into account that

$$\frac{G_+(0)T_o}{\eta} = \frac{G_-^{-1}(0)R_o}{\eta} \tag{1.5.8}$$

Now the first member of (1.5.6) is regular in $\mathrm{Im}[\eta] \geq 0$ and the second member is regular in $\mathrm{Im}[\eta] \leq 0$. This means that $w(\eta)$ is an entire function. It is important to observe that while the regularity properties of the minus and plus functions are assumed in the principal sheet, the entire function $w(\eta)$ is a uniform function that, for analytic continuation, assumes the same value in all the sheets introduced by the involved multivalued plus and minus functions.

Given that the factorization (1.5.1) is regular, (1.5.6) vanishes as $\eta \to \infty$.

According to Liouville's theorem, $w(\eta)$ must be vanishing everywhere. It yields the decoupling of equation (1.5.6) into

$$G_+(\eta)F_+(\eta) - C(\eta) = 0 \tag{1.5.9}$$

$$[G_-(\eta)]^{-1} F_-(\eta) - C(\eta) = 0 \tag{1.5.10}$$

It follows the solutions

$$F_+(\eta) = \sum_i \frac{[G_+(\eta)]^{-1} G_+(\eta_{ui}) R_{F+}^{n.s.}(\eta_{ui})}{\eta - \eta_{ui}} + \sum_i \frac{[G_+(\eta)]^{-1} G_-^{-1}(\eta_{li}) R_{F-}^{n.s.}(\eta_{li})}{\eta - \eta_{li}} + \frac{G_+(0)T_o}{\eta} \tag{1.5.11}$$

$$F_-(\eta) = \sum_i \frac{G_-(\eta)G_+(\eta_i)R_{F_+}^{n.s.}(\eta_i)}{\eta - \eta_i} + \sum_i \frac{G_-(\eta)G_-^{-1}(\eta_i)R_{F_-}^{n.s.}(\eta_i)}{\eta - \eta_i} + \frac{G_-^{-1}(0)R_o}{\eta} \tag{1.5.12}$$

In particular if the non-standard poles are with $\eta_o \neq 0$ we get:

$$F_+(\eta) = [G_+(\eta)]^{-1}\left[G_+(\eta_o)F_+^{n.s.}(\eta) + G_-^{-1}(\eta_o)F_-^{n.s.}(\eta)\right] \tag{1.5.13}$$

$$F_-(\eta) = [G(\eta)]^{-1} F_+(\eta) \tag{1.5.14}$$

As is well known in the literature, scalar kernels are always explicitly factorable; thus, the scalar case (the simplest class of WH equations) can always be solved in closed form.

Conversely, in the vector case, there is no closed-form solution for the general case. In fact, in spite of the great efforts made by many scientists, there has been little progress towards a general method of exact factorization.

The state of the art is now that in the 80 years since the solution of the scalar equation was found, the vector WH equation has been solved in closed form only in very particular cases for specific forms of the matrix $\mathbf{G}(\eta)$, and these methods often do not allow small variations of the problem itself.

A complete discussion of the matrices that can be factorized in closed form is presented in Chapter 4 of Daniele and Zich (2014). However, we must also observe that when an explicit factorization is possible, it often introduces very complicated multivalued functions that are expressed by integrals not amenable to explicit evaluation.

In the following section, we propose an alternative way to solve the WH equation (1.4.2) based on the reduction of this equation to a Fredholm integral equation (FIE) of the second kind (Fredholm factorization method). Although this method requires the solution of integral equations, the procedure is effective and efficient in solving general forms of WH equations in a semi-analytical fashion, preserving the spectral physical properties of the problem, such as in analytical closed-form solutions.

While solving the WH equations with the Fredholm factorization method, the immediate outcome of the semi-analytic procedure is that the approximate solution is an analytic element of the spectra $F_+(\eta)$ and $F_-(\eta)$ that require the analytic continuation procedure to obtain the spectra in the entire complex plane η. The analytic continuation procedure is obtained from the WH formulation of the problem.

1.6. The decomposition (Cauchy) equations

An important procedure in the WH technique is the decomposition (additive factorization) of a given function $F(\eta)$, $(F(\eta) \to 0$ as $\eta \to \infty)$ in standard plus and minus contributions. This operation can be accomplished by introducing the "smile" γ_1 and the "frown" γ_2 integration lines (Daniele and Zich 2014; Abrahams and Wickham 1988), as shown in Figure 1.6.1.

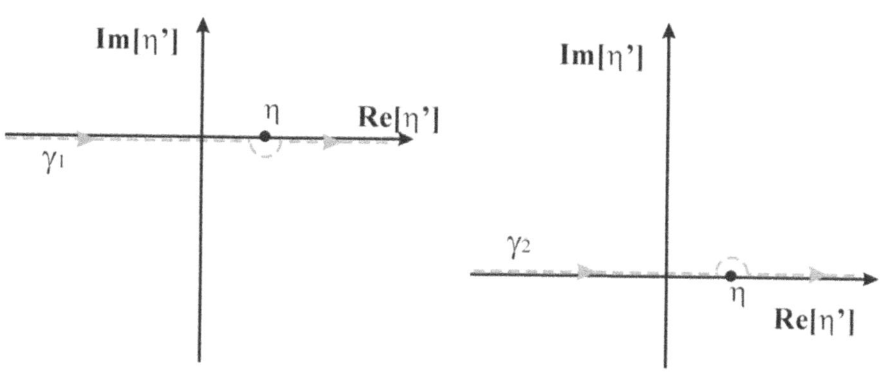

Figure 1.6.1. *The "smile" and "frown" integration lines of the η'-plane*

For example, the following equations define on the real axis of the η-plane ($\mathrm{Im}[\eta] = 0$) the standard plus and minus decomposition of $F(\eta) = F_-(\eta) + F_+(\eta)$ (Noble 1958):

$$F_+(\eta) = \frac{1}{2\pi j}\int_{\gamma_1} \frac{F(\eta')}{\eta' - \eta} d\eta' = \frac{1}{2}F(\eta) + \frac{P.V.}{2\pi j}\int_{-\infty}^{\infty} \frac{F(\eta')}{\eta' - \eta} d\eta', \ \mathrm{Im}[\eta] = 0 \quad (1.6.1)$$

$$F_-(\eta) = -\frac{1}{2\pi j}\int_{\gamma_2} \frac{F(\eta')}{\eta' - \eta} d\eta' = -\frac{1}{2}F(\eta) + \frac{P.V.}{2\pi j}\int_{-\infty}^{\infty} \frac{F(\eta')}{\eta' - \eta} d\eta', \ \mathrm{Im}[\eta] = 0 \quad (1.6.2)$$

The values of the decomposed functions (1.6.1) and (1.6.2) outside the real axis can be obtained by analytic continuation.

By applying (1.6.1) and (1.6.2) to the non-standard plus and minus functions, we obtain the generalized Cauchy formula:

$$\frac{1}{2\pi j}\int_{\gamma_1} \frac{F_+(\eta')}{\eta' - \eta} d\eta' = F_+(\eta) - F_+^{n.s.}(\eta) \qquad \eta \in \mathbb{R} \qquad (1.6.3)$$

$$\frac{1}{2\pi j}\int_{\gamma_1} \frac{F_-(\eta')}{\eta' - \eta} d\eta' = F_-^{n.s.}(\eta) \qquad \eta \in \mathbb{R} \qquad (1.6.4)$$

$$\frac{1}{2\pi j}\int_{\gamma_2} \frac{F_-(\eta')}{\eta' - \eta} d\eta' = -F_-(\eta) + F_-^{n.s.}(\eta) \qquad \eta \in \mathbb{R} \qquad (1.6.5)$$

$$\frac{1}{2\pi j}\int_{\gamma_2} \frac{F_+(\eta')}{\eta' - \eta} d\eta' = -F_+^{n.s.}(\eta) \qquad \eta \in \mathbb{R} \qquad (1.6.6)$$

In the following, we provide the proof of (1.6.3); a similar methodology can then be applied to prove (1.6.4)–(1.6.6).

Since $F_+(\eta') - F_+^{n.s.}(\eta')$ is standard, we can apply (1.6.1) to obtain

$$F_+(\eta) - F_+^{n.s.}(\eta) = \frac{1}{2\pi j}\int_{\gamma_1} \frac{F_+(\eta') - F_+^{n.s.}(\eta')}{\eta' - \eta} d\eta' \qquad (1.6.7)$$

For the sake of simplicity, let us assume the presence of only one non-standard pole; thus

$$\begin{aligned} F_+(\eta) - F_+^{n.s.}(\eta) &= \frac{1}{2\pi j}\int_{\gamma_1} \frac{F_+(\eta') - \dfrac{T_i}{\eta' - \eta_i}}{\eta' - \eta} d\eta' \\ &= \frac{1}{2\pi j}\int_{\gamma_1} \frac{F_+(\eta')}{\eta' - \eta} d\eta' - \frac{1}{2\pi j}\int_{\gamma_1} \frac{T_i}{(\eta' - \eta_i)(\eta' - \eta)} d\eta' \end{aligned} \qquad (1.6.8)$$

By closing γ_1 with a semi-circle at infinity in the lower half-plane of the complex plane η, the last term in (1.6.8) becomes zero; therefore, (1.6.3) holds.

A slight modification occurs when the non-standard pole is $\eta_i = \eta_o = 0$. In this case, the presence of the pole on the integration line γ_1 requires that the integral at the second member of (1.6.7) be evaluated as a principal value when split:

$$\begin{aligned} F_+(\eta') - F_+^{n.s.}(\eta') &= \frac{1}{2\pi j}\int_{\gamma_1} \frac{F_+(\eta') - F_+^{n.s.}(\eta')}{\eta'-\eta} d\eta' \\ &= \frac{1}{2\pi j} P.V.\int_{\gamma_1} \frac{F_+(\eta')}{\eta'-\eta} d\eta' - \frac{1}{2\pi j} P.V.\int_{\gamma_1} \frac{F_+^{n.s.}(\eta')}{\eta'-\eta} d\eta' \end{aligned} \tag{1.6.9}$$

Now we have:

$$\frac{1}{2\pi j} P.V.\int_{\gamma_1} \frac{F_+^{n.s.}(\eta')}{\eta'-\eta} d\eta' = \frac{1}{2\pi j}\int_{\bar{\gamma}_1} \frac{F_+^{n.s.}(\eta')}{\eta'-\eta} d\eta' - \frac{1}{2} F_+^{n.s.}(\eta') = -\frac{1}{2} F_+^{n.s.}(\eta') \tag{1.6.10}$$

where $\bar{\gamma}_1$ is the modified smile line that is warped such that the pole $\eta_i = \eta_o = 0$ is located above of this line. By closing $\bar{\gamma}_1$ with a semi-circle at infinity in the lower half-plane of the complex plane η, the integral becomes zero; thus, in this case, (1.6.3) becomes

$$\frac{1}{2\pi j} P.V.\int_{\gamma_1} \frac{F_+(\eta')}{\eta'-\eta} d\eta' = F_+(\eta') - \frac{1}{2} F_+^{n.s.}(\eta') \tag{1.6.11}$$

If we apply the Cauchy decomposition formula in section 1.5, it yields the following relations for (1.5.7):

$$C_+(\eta) = \frac{1}{2\pi j}\int_{\gamma_1} \frac{[G_-(\eta')]^{-1} F_-^{GOn.s.}(\eta')}{\eta'-\eta} d\eta' \tag{1.6.12}$$

$$C_-(\eta) = -\frac{1}{2\pi j}\int_{\gamma_2} \frac{G_+(\eta') F_+^{GOn.s.}(\eta')}{\eta'-\eta} d\eta' \tag{1.6.13}$$

1.7. General formula for factorization of scalar kernels

The fundamental idea for factorizing (multiplicative factorization) a given function $G(\eta)$ is based on the logarithmic decomposition. By introducing

$$G(\eta) = e^{Log[G(\eta)]} = e^{\psi(\eta)} \tag{1.7.1}$$

the decomposition of $\psi(\eta) = \log[G(\eta)]$ yields

$$\log[G(\eta)] = \psi(\eta) = \psi_-(\eta) + \psi_+(\eta) \tag{1.7.2}$$

with:

$$\psi_+(\eta) = \frac{1}{2\pi j}\int_{\gamma 1} \frac{\log G(\eta')]}{\eta' - \eta} d\eta \tag{1.7.3}$$

$$\psi_-(\eta) = \psi(\eta) - \psi_+(\eta) \tag{1.7.4}$$

where γ_1 is the "smile" real axis (Figure 1.6.1).

By taking into account that in the scalar case:

$$e^{\psi(\eta)} = e^{\psi_-(\eta)+\psi_+(\eta)} = e^{\psi_-(\eta)} e^{\psi_+(\eta)} \tag{1.7.5}$$

we obtain the factorization formula

$$G(\eta) = G_-(\eta) G_+(\eta) \tag{1.7.6}$$

where:

$$G_+(\eta) = \exp\left[\frac{1}{2\pi j}\int_{\gamma_1} \frac{\log\left[G(\eta')\right]}{\eta' - \eta} d\eta\right] \tag{1.7.7}$$

$$G_-(\eta) = G(\eta) G_+^{-1}(\eta) \tag{1.7.8}$$

If $G(\eta)$ is even, we obtain the property and simplification:

$$G_-(\eta) = G_+(-\eta) \tag{1.7.9}$$

The integral that defines $G_+(\eta)$ in (1.7.7) converges if $G(\eta) \to 1$, as $\eta \to \infty$, but most often this property does not occur.

In these cases, the idea is usually to normalize $G(\eta) = f(\eta)K(\eta)$ such that $K(\eta) \to 1$, as $\eta \to \infty$, and $f(\eta)$ may be factorized with other techniques.

1.8. Some explicit factorization of matrix kernels useful in wedge scattering

In this section, we illustrate some important closed-form exact and explicit factorization for matrix kernels.

First, (1.7.7) is also valid for matrix kernels, provided that the matrices $\psi_-(\eta)$ and $\psi_+(\eta)$ do commute. Unfortunately, $\psi_-(\eta)$ and $\psi_+(\eta)$ only commute in very few special cases.

In wedge scattering problems, we often deal with the factorization of matrices of order 2 having the form:

$$G(\eta) = \begin{vmatrix} 1 & a(\eta) \\ b(\eta) & 1 \end{vmatrix} \tag{1.8.1}$$

where $\dfrac{a(\eta)}{b(\eta)} = \dfrac{n(\eta)}{p(\eta)}$ with polynomial $n(\eta)$ and $p(\eta)$.

These matrices are particular cases of the ones discussed, in general, in Daniele and Zich (2014). They can be factorized explicitly by using the equations

$$G(\eta) = \begin{vmatrix} 1 & a(\eta) \\ b(\eta) & 1 \end{vmatrix} = \hat{G}_-(\eta)\hat{G}_+(\eta) \tag{1.8.2}$$

where:

$$\hat{G}_-(\eta) = \sqrt{g_-(\eta)}\exp[\frac{1}{2}t_-(\eta)\begin{vmatrix} 0 & n(\eta) \\ p(\eta) & 0 \end{vmatrix}] =$$

$$= \sqrt{g_-(\eta)}\begin{vmatrix} \cosh[\frac{1}{2}t_-(\eta)\sqrt{n(\eta)p(\eta)}] & \sqrt{\frac{n(\eta)}{p(\eta)}}\sinh[\frac{1}{2}t_-(\eta)\sqrt{n(\eta)p(\eta)}] \\ \sqrt{\frac{p(\eta)}{n(\eta)}}\sinh[\frac{1}{2}t_-(\eta)\sqrt{n(\eta)p(\alpha)}] & \cosh[\frac{1}{2}t_-(\eta)\sqrt{n(\eta)p(\eta)}] \end{vmatrix} \tag{1.8.3}$$

$$\hat{G}_+(\eta) = \sqrt{g_+(\eta)}\,\exp[\frac{1}{2}t_+(\eta)\begin{vmatrix} 0 & n(\eta) \\ p(\eta) & 0 \end{vmatrix}] =$$

$$\hat{G}_+(\eta) = \sqrt{g_+(\eta)}\begin{vmatrix} \cosh[\frac{1}{2}t_+(\eta)\sqrt{n(\eta)p(\eta)}] & \sqrt{\frac{n(\eta)}{p(\eta)}}\sinh[\frac{1}{2}t_+(\eta)\sqrt{n(\eta)p(\eta)}] \\ \sqrt{\frac{p(\eta)}{n(\eta)}}\sinh[\frac{1}{2}t_+(\eta)\sqrt{n(\eta)p(\eta)}] & \cosh[\frac{1}{2}t_+(\eta)\sqrt{n(\eta)p(\eta)}] \end{vmatrix} \quad (1.8.4)$$

In the above expressions, $g_-(\eta)$ and $g_+(\eta)$ are the factorized scalars of the determinant of $G(\eta)$:

$$g(\eta) = Det[G(\eta)] = 1 - a(\eta)b(\eta) = g_-(\eta)g_+(\eta) \quad (1.8.5)$$

The functions $t_-(\eta)$ and $t_+(\eta)$ follow from the decomposition of the function $t(\eta)$ defined by:

$$t(\eta) = \frac{1}{\sqrt{n(\eta)p(\eta)}}\log\frac{1+\sqrt{a(\eta)b(\eta)}}{1-\sqrt{a(\eta)b(\eta)}} = \frac{1}{\sqrt{n(\eta)p(\eta)}}\log\frac{g(\eta)}{\left[1-\sqrt{a(\eta)b(\eta)}\right]^2} \quad (1.8.6)$$
$$= t_-(\eta) + t_+(\eta)$$

with

$$t_+(\eta) = \frac{1}{2\pi j}\int_{\gamma_1}\frac{t(u)}{u-\eta}du \quad (1.8.7)$$

$$t_-(\eta) = \frac{1}{2\pi j}\int_{\gamma_2}\frac{t(u)}{u-\eta}du \quad (1.8.8)$$

where γ_1 and γ_2 are respectively the "smile" and the "frown" real axis (Figure 1.6.1).

From (1.8.6), the singularities of $t(\eta)$ are the same as that of the $g(\eta)$ function, for example when $g(\eta)$ is null, infinity or shows branch points; therefore, the introduction of $t(\eta)$ does not add any further singularities.

The same technique can be extended to particular matrices of order 4.

Starting from the identity

$$\begin{vmatrix} 1 & a \\ b & 1 \end{vmatrix} = \sqrt{1-ab}\begin{vmatrix} \cosh\left(\frac{1}{2}\log\frac{1+\sqrt{ab}}{1-\sqrt{ab}}\right) & \sqrt{\frac{a}{b}}\sinh\left(\frac{1}{2}\log\frac{1+\sqrt{ab}}{1-\sqrt{ab}}\right) \\ \sqrt{\frac{b}{a}}\sinh\left(\frac{1}{2}\log\frac{1+\sqrt{ab}}{1-\sqrt{ab}}\right) & \cosh\left(\frac{1}{2}\log\frac{1+\sqrt{ab}}{1-\sqrt{ab}}\right) \end{vmatrix} \tag{1.8.9}$$

that holds when a and b are scalar and also when $a = \overline{a}\, f$, $b = \overline{b}\, f$ are matrices of order 2 with scalar $\overline{a}, \overline{b}$, if $\frac{\overline{a}}{\overline{b}} = \frac{n}{p}$ with polynomial n and p, it is possible to rewrite (1.8.9) in the form:

$$\begin{vmatrix} 1 & a \\ b & 1 \end{vmatrix} = \sqrt{1-\overline{a}\overline{b}\, f^2}\begin{vmatrix} \cosh\left(\frac{1}{2}\log\frac{1+f\sqrt{\overline{a}\overline{b}}}{1-f\sqrt{\overline{a}\overline{b}}}\right) & \sqrt{\frac{n}{p}}\sinh\left(\frac{1}{2}\log\frac{1+f\sqrt{\overline{a}\overline{b}}}{1-f\sqrt{\overline{a}\overline{b}}}\right) \\ \sqrt{\frac{p}{n}}\sinh\left(\frac{1}{2}\log\frac{1+f\sqrt{\overline{a}\overline{b}}}{1-f\sqrt{\overline{a}\overline{b}}}\right) & \cosh\left(\frac{1}{2}\log\frac{1+f\sqrt{\overline{a}\overline{b}}}{1-f\sqrt{\overline{a}\overline{b}}}\right) \end{vmatrix} \tag{1.8.10}$$

If f is diagonal, a commutative algebra is involved, and it is possible to manipulate the matrices as scalars. For example, we again obtain the factorized matrices (1.8.3) and (1.8.4), where the diagonal matrices $g_{-,+}$ and $t_{-,+}$ are defined by

$$g = \sqrt{1-\overline{a}\overline{b}\, f^2} = g_- g_+ \tag{1.8.11}$$

$$t = \frac{1}{\sqrt{np}}\log\frac{1+f\sqrt{\overline{a}\overline{b}}}{1-f\sqrt{\overline{a}\overline{b}}} = t_- + t_+ \tag{1.8.12}$$

Other explicit factorizations are possible when the eigenvectors of f are constant. These factorizations will be used in Chapter 4. When f presents more general forms, in several cases, explicit factorizations are still possible. However, for brevity, they will not be presented in this book as well as the general case of matrices (1.8.9) of arbitrary order.

In section 4.5.1, we will deal with the problem of scattering by impenetrable wedge whose WH equations have the form:

$$\begin{vmatrix} \xi 1 & 1 \\ \xi 1 & -1 \end{vmatrix} X_+ = \begin{vmatrix} D_a & 0 \\ 0 & D_b \end{vmatrix} Y_- \tag{1.8.13}$$

where the matrices $D_a = D_a(\eta)$ and $D_b(\eta)$ are diagonal matrices of order 2, $\xi = \xi(\eta)$ is a scalar function and 1 is the identity matrix of order 2.

By inverting the matrix in the second member of (1.8.13), it yields:

$$\begin{vmatrix} \xi D_a^{-1} & D_a^{-1} \\ \xi D_b^{-1} & -D_b^{-1} \end{vmatrix} X_+ = Y_- \tag{1.8.14}$$

We now have:

$$\begin{vmatrix} \xi D_a^{-1} & D_a^{-1} \\ \xi D_b^{-1} & -D_b^{-1} \end{vmatrix} = \frac{1}{2} \begin{vmatrix} 1 & 1 \\ 1 & -1 \end{vmatrix} \begin{vmatrix} \xi D_\Sigma & D_\Delta \\ \xi D_\Delta & D_\Sigma \end{vmatrix} \tag{1.8.15}$$

where:

$$D_\Sigma = \frac{1}{2}(D_A^{-1} + D_B^{-1}) \tag{1.8.16}$$

$$D_\Delta = \frac{1}{2}(D_A^{-1} - D_B^{-1}) \tag{1.8.17}$$

Since the involved sub-matrices of order 2 are all diagonal, they commute so that the factorization of $D_\Sigma = D_{\Sigma-} D_{\Sigma+}$ and $D_\Delta = D_{\Delta-} D_{\Delta+}$ allows the following reduction:

$$\begin{vmatrix} \xi D_\Sigma & D_\Delta \\ \xi D_\Delta & D_\Sigma \end{vmatrix} = \begin{vmatrix} \xi_- D_{\Sigma-} & 0 \\ 0 & D_{\Sigma-} \end{vmatrix} \begin{vmatrix} 1 & \frac{1}{\xi_-} D_{\Sigma-}^{-1} D_\Delta D_{\Sigma+}^{-1} \\ \xi_- D_{\Sigma-}^{-1} D_\Delta D_{\Sigma+}^{-1} & 1 \end{vmatrix} \begin{vmatrix} \xi_+ D_{\Sigma+} & 0 \\ 0 & D_{\Sigma+} \end{vmatrix} \tag{1.8.18}$$

Taking into account that the factorization of $D_\Sigma = D_{\Sigma-} D_{\Sigma+}$ and $D_\Delta = D_{\Delta-} D_{\Delta+}$ involves only factorization of scalars, the previous expression reduces the problem to the factorization of the matrix of order 4, which has the form (1.8.10):

$$G_d = \begin{vmatrix} 1 & \dfrac{1}{\xi_-} f \\ \xi_- f & 1 \end{vmatrix} = \begin{vmatrix} 1 & \overline{a} f \\ \overline{b}\, f & 1 \end{vmatrix} \tag{1.8.19}$$

Since the matrix of order 2 $f = D_{\Sigma-}^{-1} D_{\Delta} D_{\Sigma+}^{-1}$ is diagonal and in impenetrable wedge problems ξ_-^2 is a polynomial of first degree (Chapter 3) ($\overline{a} = \dfrac{1}{\xi_-}$, $\overline{b} = \xi_-$), the matrix (1.8.19) has the form that is amenable to closed-form factorization.

The factorized matrices are provided by the extension of (1.8.3) and (1.8.4) to the case of the matrix kernel (1.8.10).

An updated theory of the matrices amenable to exact factorization is presented in Chapter 4 of Daniele and Zich (2014). Beyond matrices having particular forms, an important class of them consists of the matrices that commute with polynomial matrices. For these matrices, the logarithmic decomposition can be applied. However, excluding the cases of very simple polynomial matrices, in general, the logarithmic decomposition introduces problems with essential singularities at infinity, which must be eliminated. A theory for the elimination of the essential singularities has been presented in Daniele and Zich (2014). However, with the exception of important cases, it is very cumbersome and, in practice, it is better to use alternative techniques if available.

1.9. The Fredholm factorization technique: a general technique to solve WH equations

According to these authors, the most powerful approximate and general method to solve WH equations is based on the so-called Fredholm factorization (Daniele and Lombardi 2007, Daniele and Zich 2014). An introduction to this factorization is proposed in this section with an application to the electromagnetic scattering by a half-plane problem, discussed in the following chapter.

In WH equations, plus and minus functions have the same importance. However, in order to solve the equations, we must try to separate their role in order to have two different equations involving only one of kind of unknowns. As with the classical factorization described in the previous sections, the Fredholm factorization technique is also based on this idea.

To be concise, henceforth we will focus on the determination of the plus function, but it can be generalized easily to the direct determination of the minus function by following similar steps.

As discussed in the previous sections, the classical factorization scheme "eliminates" by separating the minus function $F_-(\eta)$ through the factorization of the kernel $G(\eta)$.

The Fredholm factorization method alternatively starts from the elimination of the minus function using the Cauchy decomposition formulas presented in section 1.6.

Taking into account the homogeneous WH equation

$$G(\eta)F_+(\eta) = F_-(\eta) \tag{1.9.1}$$

by applying the decomposition formulas, we obtain:

$$\frac{1}{2\pi j}\int_{\gamma_1}\frac{G(\eta')F_+(\eta')}{\eta'-\eta}d\eta' = \frac{1}{2}G(\eta)F_+(\eta) + \frac{P.V.}{2\pi j}\int_{-\infty}^{\infty}\frac{G(\eta')F_+(\eta')}{\eta'-\eta}d\eta' \tag{1.9.2}$$

$$\frac{1}{2\pi j}\int_{\gamma_1}\frac{F_-(\eta')}{\eta'-\eta}d\eta' = F_-^{GOn.s.}(\eta) \tag{1.9.3}$$

$$\frac{1}{2\pi j}\int_{\gamma_2}\frac{G(\eta)F_+(\eta')}{\eta'-\eta}d\eta' = -\frac{1}{2}G(\eta)F_+(\eta) + \frac{P.V.}{2\pi j}\int_{-\infty}^{\infty}\frac{G(\eta)F_+(\eta')}{\eta'-\eta}d\eta' \tag{1.9.4}$$

$$\frac{1}{2\pi j}\int_{\gamma_2}\frac{G(\eta)F_+(\eta')}{\eta'-\eta}d\eta' = -G(\eta)F_+^{n.s.}(\eta) = -G(\eta)F_+^{GOn.s.}(\eta) \tag{1.9.5}$$

where $\mathrm{Im}[\eta]=0$ and γ_1 (γ_2) is the smile (frown) integration line (Figure 1.6.1).

In the procedure, we first apply the smile contour integration γ_1 to (1.9.1), yielding (1.9.2) for the LHS and (1.9.3) for the RHS, and we then subtract on both sides of the equation the same term considering the form (1.9.4) on the LHS and the equivalent (1.9.5) on the RHS. We obtain

$$G(\eta)F_+(\eta) + \frac{1}{2\pi j}\int_{-\infty}^{\infty}\frac{[G(\eta')-G(\eta)]F_+(\eta')}{\eta'-\eta}d\eta' = F_-^{GOn.s.}(\eta) + G(\eta)F_+^{GOn.s.}(\eta) \tag{1.9.6}$$

or equivalently $F_+(\eta) = -\frac{[G(\eta)]^{-1}}{2\pi j}\int_{-\infty}^{\infty}\frac{[G(\eta')-G(\eta)]F_+(\eta')}{\eta'-\eta}d\eta' + F_o(\eta)$ where $\mathrm{Im}[\eta]=0$ and

$$F_o(\eta) = [G(\eta)]^{-1} F_-^{GOn.s.}(\eta) + F_+^{GOn.s.}(\eta) \tag{1.9.7}$$

is known. Equation (1.9.6) does not contain the minus unknown $F_-(\eta)$.

We assert that, due to the construction procedure, (1.9.6) is always a Fredholm integral equation of the second kind (FIE) for the unknown $F_+(\eta)$ (see also Appendix 1.B).

A remarkable property is that the integral in (1.9.6) does not present the singularities of $F_+(\eta)$ (Daniele and Zich 2014), but only models correction diffractive terms.

Explicit solutions of (1.9.6) can be obtained if the closed-form factorization of $G(\eta) = G_-(\eta)G_+(\eta)$ is available. Without loss of generality, we show this property when $F_+(\eta)$ is the standard (i.e. $F_+^{GOn.s.}(\eta) = 0 \rightarrow C_-(\eta) = 0$ in section 1.5). Assuming $X_+(\eta) = G_+(\eta)F_+(\eta)$, after some mathematical manipulations, (1.9.6) becomes:

$$\begin{aligned} & X_+(\eta) + \frac{1}{2\pi j}\int_{-\infty}^{\infty} \frac{\left[[G_-(\eta)]^{-1} G_-(\eta') - G_+(\eta)[G_+(\eta')]^{-1}\right] X_+(\eta')}{\eta' - \eta} d\eta' \\ & = [G_-(\eta)]^{-1} F_-^{GOn.s.}(\eta) \end{aligned} \tag{1.9.8}$$

Now considering the decompositions formulas of section 1.6:

$$\begin{aligned} & \frac{1}{2\pi j}\int_{-\infty}^{\infty} \frac{\left[[G_-(\eta)]^{-1} G_-(\eta') - G_+(\eta)[G_+(\eta')]^{-1}\right] X_+(\eta')}{\eta' - \eta} d\eta' = \\ & \frac{P.V.}{2\pi j}\int_{-\infty}^{\infty} \frac{[G_-(\eta)]^{-1} G_-(\eta') X_+(\eta')}{\eta' - \eta} d\eta' - \frac{P.V.}{2\pi j}\int_{-\infty}^{\infty} \frac{G_+(\eta)[G_+(\eta')]^{-1} X_+(\eta')}{\eta' - \eta} d\eta' = \\ & \frac{1}{2\pi j}\int_{\gamma_2} \frac{[G_-(\eta)]^{-1} G_-(\eta') X_+(\eta')}{\eta' - \eta} d\eta' + \frac{1}{2} X_+(\eta) - X_+(\eta) + \frac{1}{2} X_+(\eta) = \\ & = \frac{1}{2\pi j}\int_{\gamma_2} \frac{[G_-(\eta)]^{-1} G_-(\eta') X_+(\eta')}{\eta' - \eta} d\eta' = Y_-(\eta) \end{aligned} \tag{1.9.9}$$

where $Y_-(\eta)$ is a standard minus function, it yields:

$$X_+(\eta) + Y_-(\eta) = [G_-(\eta)]^{-1} F_-^{GOn.s.}(\eta) \tag{1.9.10}$$

The smile integration of (1.9.10) yields:

$$X_+(\eta) = \frac{1}{2\pi j}\int_{\gamma_1} \frac{[G_-(\eta')]^{-1} F_-^{GOn.s.}(\eta')}{\eta'-\eta} d\eta' \tag{1.9.11}$$

Taking into account (1.6.12) and (1.9.11), i.e. $X_+(\eta) = C_+(\eta)$, we obtain the same solution of section 1.5 for (1.5.9) with $C_-(\eta) = 0$.

We note that the Fredholm factorization method is applicable to general vector WH equations (1.9.1).

1.10. Spectral properties of the unknowns

The singularities of WH unknowns and, in general, of spectral functions are very important from a mathematical and physical point of view. For example, they produce GO plane waves, surface waves, lateral waves, leaky waves and so on. As in closed-form factorization, the Fredholm factorization provides the location of the singularities of the unknowns, with the help of analytic continuation. Let us focus on $F_+(\eta)$ of (1.9.6):

$$F_+(\eta) = -\frac{[G(\eta)]^{-1}}{2\pi j}\int_{-\infty}^{\infty} \frac{[G(\eta') - G(\eta)]F_+(\eta')}{\eta'-\eta} d\eta' + F_o(\eta) \tag{1.10.1}$$

with $F_o(\eta) = [G(\eta)]^{-1} F_-^{GOn.s.}(\eta) + F_+^{GOn.s.}(\eta)$.

Moving the observation point η from the integration line where $F_+(\eta)$ is regular, an analytic continuation of equation (1.10.1) allows us to correctly verify all the singularities of $F_+(\eta)$. In particular, since the integral does not involve the singularities of $F_+(\eta)$ see section 5.1.4 of (Daniele and Zich 2014), the GO poles of $F_+(\eta)$ derive only from those of $F_o(\eta)$. Without loss of generality, henceforth, we will only consider the presence of the GO pole η_o. From (1.9.7), we have:

$$F_o(\eta) = [G(\eta_o)]^{-1} F_-^{GOn.s.}(\eta) + F_+^{GOn.s.}(\eta) + ([G(\eta)]^{-1} - [G(\eta_o)]^{-1})F_-^{GOn.s.}(\eta) \tag{1.10.2}$$

Now the first and third terms in (1.10.2) are null if $F_-^{GO}(\eta)$ is standard, thus $F_o(\eta) = F_+^{GOn.s.}(\eta) = F_+^{GO}(\eta)$. Conversely, the second term $F_+^{GOn.s.}(\eta)$ of (1.10.2) is null if $F_+^{GO}(\eta)$ is standard. In the latter case from (1.9.1), we have $[G(\eta_o)]^{-1})F_-^{GOn.s.}(\eta) = F_+^{GO}(\eta)$ thus $F_o(\eta) = [G(\eta)]^{-1} F_-^{GOn.s.}(\eta) = F_+^{GO}(\eta) + ([G(\eta)]^{-1} - [G(\eta_o)]^{-1})F_-^{GOn.s.}(\eta)$.

Therefore, in general we have

$$F_o(\eta) = F_+^{GO}(\eta) + ([G(\eta)]^{-1} - [G(\eta_o)]^{-1})F_-^{GOn.s.}(\eta) \qquad (1.10.3)$$

Note that the last term in (1.10.3) is a contribution with an apparent pole η_o.

By substituting (1.10.3) in (1.10.1), we prove that $F_+(\eta) - F_+^{GO}(\eta)$ does not contain GO singularities (poles).

Beyond the GO poles whose contributions are known, $F_+(\eta)$ may contain structural singularities due to the kernel. Looking at (1.10.1), we need to consider the singularities of $[G(\eta)]^{-1}$ and $G(\eta)$.

It is remarkable that the singularities of $[G(\eta)]$ do not appear in $F_+(\eta)$ and only the singularities of $[G(\eta)]^{-1}$ located in the lower half-plane $\mathrm{Im}[\eta] < 0$ has contributions in $F_+(\eta)$.

Concerning the effects of the singularities of $G(\eta)$ on $F_+(\eta)$, we consider, for the sake of simplicity, only the case where the structural singularities are simple poles. Let η_r be a pole of $G(\eta)$.

We put:

$$G(\eta) = \frac{R(\eta_r)}{\eta - \eta_r} + h(\eta), \quad G(\eta') = \frac{R(\eta_r)}{\eta' - \eta_r} + h(\eta') \qquad (1.10.4)$$

where $R(\eta_r)$ is the residue of $G(\eta)$ in η_r and $h(\eta)$ is regular at η_r.

By substituting (1.10.4) into the integral of (1.10.1), we obtain

$$\begin{aligned}\int_{-\infty}^{\infty} \frac{G(\eta') - G(\eta)}{\eta' - \eta} F_+(\eta') d\eta' &= -\int_{-\infty}^{\infty} \frac{R(\eta_r)}{(\eta' - \eta_r)(\eta - \eta_r)} F_+(\eta') d\eta' \\ &+ \int_{-\infty}^{\infty} \frac{h(\eta') - h(\eta)}{(\eta' - \eta)} F_+(\eta') d\eta'\end{aligned} \qquad (1.10.5)$$

We assert that this contribution is regular at η_r when either $\mathrm{Im}[\eta_r]>0$ or $\mathrm{Im}[\eta_r]<0$. Starting from $\mathrm{Im}[\eta_r]>0$, we obtain

$$\begin{aligned}&-\frac{1}{2\pi j}\int_{-\infty}^{\infty}\frac{[G(\eta)]^{-1}R(\eta_r)}{(\eta'-\eta_r)(\eta-\eta_r)}F_+(\eta')d\eta'+[G(\eta)]^{-1}\int_{-\infty}^{\infty}\frac{h(\eta')-h(\eta)}{(\eta'-\eta)}F_+(\eta')d\eta'=\\&-\frac{[G(\eta)]^{-1}R(\eta_r)}{(\eta-\eta_r)}F_+(\eta_r)+[G(\eta)]^{-1}\int_{-\infty}^{\infty}\frac{h(\eta')-h(\eta)}{(\eta'-\eta)}F_+(\eta')d\eta'\end{aligned}\qquad(1.10.6)$$

By considering from (1.10.4) that $[G(\eta)]^{-1}(\frac{R(\eta_r)}{\eta-\eta_r}+h(\eta))=\mathbf{1}$, we obtain $[G(\eta_r)]^{-1}R(\eta_r)=0$ (we recall that η_r is the pole of only $G(\eta)$, not $G^{-1}(\eta)$); therefore, the first contribution in (1.10.6) is regular, together with the second one.

The same reasoning applies when $\mathrm{Im}[\eta_r]<0$. For example, in this case, we have directly that $\frac{1}{2\pi j}\int_{-\infty}^{\infty}\frac{R(\eta_r)}{(\eta'-\eta_r)(\eta-\eta_r)}F_+(\eta')d\eta'=0$ and in the scalar case we have that the term with integral in (1.10.1) is vanishing since $[G(\eta_r)]^{-1}=0$.

Concerning the effects of the singularities of $G^{-1}(\eta)$ on $F_+(\eta)$, we consider, for the sake of simplicity, only the case where the structural singularities are simple poles. Let η_q be a pole of $G^{-1}(\eta)$.

By considering (1.10.1), if $\mathrm{Im}[\eta_q]>0$, we have that

$$\int_{-\infty}^{\infty}\frac{[G(\eta')-G(\eta_q)]F_+(\eta')}{\eta'-\eta_q}d\eta'=\int_{-\infty}^{\infty}\frac{[G(\eta')]F_+(\eta')}{\eta'-\eta_q}d\eta'=\int_{-\infty}^{\infty}\frac{F_-(\eta')}{\eta'-\eta_q}d\eta'=0\qquad(1.10.7)$$

by applying the residue theorem and closing the contour on the lower half-plane (we recall that η_q is the pole of $G^{-1}(\eta)$, thus a zero of $G(\eta)$). From (1.10.1), we have that $F_+(\eta)$ is regular at η_q.

Conversely, if $\mathrm{Im}[\eta_q]<0$, we have that

$$\int_{-\infty}^{\infty}\frac{\left[G(\eta')-G(\eta_q)\right]F_+(\eta')}{\eta'-\eta_q}d\eta'=\int_{-\infty}^{\infty}\frac{G(\eta')F_+(\eta')}{\eta'-\eta_q}d\eta'=\int_{-\infty}^{\infty}\frac{F_-(\eta')}{\eta'-\eta_q}d\eta'=F_-(\eta_q)\neq 0 \tag{1.10.8}$$

since the pole $\text{Im}[\eta_q]<0$ is now captured when closing the contour in the lower half-plane.

From (1.10.1), we have that $F_+(\eta)$ is not regular at η_q due to (1.10.8).

Finally, we observe that beyond the GO and structural singularities, the analytic continuation of (1.10.1) may present other singularities arising from the presence of multivalued kernel $G(\eta)$.

In fact, we note that, since the line of integration is located on a Riemann sheet of $G(\eta')$ (typically the proper one), $G(\eta')-G(\eta)$ is vanishing as $\eta'\to\eta$ only if $G(\eta)$ is evaluated on the same sheet. This is not the case if the observation point η is in an improper sheet while the integration line is located in the proper sheet.

1.11. Semi-analytical solution of the Fredholm integral equation

General approximate and numerical methods for solving integral equations are well known in the literature. In particular, we recall that the integral equations derived from WH equations are Fredholm integral equations (FIEs) of the second kind (see section 1.8 and Appendix 1.B).

Since (1.9.6) is an FIE, it is a well-conditioned equation (Kantorovich and Krylov 1964); efficient techniques to solve these equations are available in the literature: for the sake of simplicity, see the Nystrom method, the method of moments and the recursive solution.

In general, the numerical solution of the FIE of the second kind is very efficient even using simple numerical quadrature.

Due to the high convergence rate of the equations, we apply a simplified form of discretization based on the sample-and-hold technique, i.e. the piecewise-constant collocation method.

However, we observed a slow convergence and degradation of numerical precision when dealing with equations where the singularities of the source and/or the structure were critical, i.e. poles and branches near the integration path. This is a well-known problem in the literature (see, for example, Kantorovich and Krylov 1964). There are several techniques to overcome this problem: for example, we can apply specialized numerical quadratures. As in Daniele and Lombardi (2007) and Daniele and Zich (2014), we suggest a different approach that is based on warping the contour path from the real-axis R into a new contour (e.g. B), being far enough from the critical singularities. Henceforth, we will consider $B-R+\Gamma$ as a closed line, where Γ is the closure often obtained with arcs located at infinity where the integrand is vanishing.

We will show that (1.9.6) is an integral equation that also holds along a contour B that is different from the integration line R for the observation point on the same line B:

$$F_+(\eta) = -\frac{[G(\eta)]^{-1}}{2\pi j}\int_B \frac{[G(\eta')-G(\eta)]F_+(\eta')}{\eta'-\eta}d\eta' + F_o(\eta) \qquad \eta \in B \qquad (1.11.1)$$

To obtain an extension of (1.9.6) and (1.11.1) for the observation point and the integration point on different lines, for example on B and R, respectively, we must deform the integration line from B to R using the Cauchy formula:

$$\int_B \frac{[G(\eta')-G(\eta)]F_+(\eta')}{\eta'-\eta}d\eta' = \int_{-\infty}^{\infty} \frac{[G(\eta')-G(\eta)]F_+(\eta')}{\eta'-\eta}d\eta' + \mathrm{f}(\eta) \qquad \eta \in \mathrm{B} \qquad (1.11.2)$$

where $\mathrm{f}(\eta)$ is the contribution of the singularity of the integrand $\frac{[G(\eta')-G(\eta)]F_+(\eta')}{\eta'-\eta}$ in the region Ω of the plane η' enclosed by the lines $B-R+\Gamma$. The singularities of the integrand can be due to the isolated singularity of $G(\eta')$ and/or $F_+(\eta')$. Furthermore, by changing the values of η on B, lines of singularities are possible for values of $\eta' \in \Omega$ such that the kernel is singular. This last possibility occurs, in particular, for generalized WH equations.

It yields

$$F_+(\eta) = -\frac{[G(\eta)]^{-1}}{2\pi j}\left(\int_{-\infty}^{\infty} \frac{[G(\eta')-G(\eta)]F_+(\eta')}{\eta'-\eta}d\eta' + \mathrm{f}(\eta)\right) + F_o(\eta) \qquad \eta \in B \ (1.11.3)$$

If possible, from a practical point of view, it is useful to avoid the contribution of singularity lines to the integral formulation, by appropriately selecting B, Γ and Ω.

We observe that in the presence of isotropic regions $G(\eta')$ and/or $F_+(\eta')$ show singularities that are only located in the second and fourth quadrants of η'. For this reason, we select B lines located in the first and third quadrants of the proper sheet of η'. With this choice, f(η)=0 and equation (1.11.1) is valid.

The selection of contours B to enhance convergence is extensively discussed in Daniele and Lombardi (2007) and Daniele and Zich (2014) (see also section 1.15). For example, a practical and appropriate line is $B \equiv B_\theta$:

$$B_\theta : \lambda_\theta(y) = e^{j\theta} y, \quad -\infty < y < \infty \text{ where } 0 < \theta \leq \pi/2 \tag{1.11.4}$$

Taking into account that B_θ defines a region Ω where η' does not produce singularities of the integrand when $\eta \in B_\theta$, we obtain a new FIE where the original integration line R is warped into the integration line B_θ:

$$F_+(\eta) = -\frac{[G(\eta)]^{-1}}{2\pi j}\int_{B_\theta} \frac{[G(\eta') - G(\eta)]F_+(\eta')}{\eta' - \eta} d\eta' + F_o(\eta) \qquad \eta \in B_\theta \tag{1.11.5}$$

Assuming in (1.11.5) that the observation point η is always on the integration line, the sampled approximated version of the equation is:

$$\begin{aligned} &G(\lambda(hn)F_{a+}(\lambda(hn)) + \frac{h}{2\pi j}\sum_{i=-A/h}^{A/h} m(\lambda(hn), \lambda(hi))F_{a+}(\lambda(hi))\lambda'(hi) \\ &= F_-^{GOn.s.}(\lambda(hn)) + G(\lambda(hn))F_+^{GOn.s.}(\lambda(hn)) \end{aligned} \tag{1.11.6}$$

with $n \in [-A/h, A/h]$ (n,i integer), where:

$$m(\eta, \eta') = If[\eta = \eta', G'(\eta), \frac{G(\eta') - G(\eta))}{\eta' - \eta}] \tag{1.11.7}$$

In (1.11.6), the quantities h and $\pm A$ respectively represent the discretization step and the endpoints of the truncated integration line.

By defining the matrices M and D of order 2A+1 respectively with non-null elements

$$M_{ni} = \frac{h}{2\pi j} m[\lambda(hn), t(hi)], \quad n \in [-A/h, A/h], i \in [-A/h, A/h] \tag{1.11.8}$$

$$D_{ii} = G[\lambda(hi)], \quad i \in [-A/h, A/h] \tag{1.11.9}$$

and the vector s with elements

$$s_i = F_-^{GOn.s.}(\lambda(hi)) + G(\lambda(hi))F_+^{GOn.s.}(\lambda(hi)), \quad i \in [-A/h, A/h] \tag{1.11.10}$$

The solution f of the linear system

$$[D+M]f = s \tag{1.11.11}$$

provides the samples of $F_{a+}(\eta)$

$$f_i = F_{a+}(\lambda(hi)), \quad i \in [-A/h, A/h] \tag{1.11.12}$$

It yields the approximate solution:

$$\begin{aligned} F_{a+}(\eta) = G^{-1}(\eta)\left[-\frac{h}{2\pi j}\cdot\sum_{i=-A/h}^{A/h} m(\eta,\lambda(hi))F_{a+}(\lambda(hi))\lambda'(hi)\right] \\ +G^{-1}(\eta)F_-^{GOn.s.}(\eta) + F_+^{GOn.s.}(\eta) \end{aligned} \tag{1.11.13}$$

From theoretical considerations (Kantorovich and Krylov 1964), we ascertain that by reducing h and increasing A the numerical solution converges to the exact solution of the integral equation.

We observe that a local loss of precision in (1.11.13) is for the presence of additional spurious singularities due to the approximation of the integral with a finite sum, as discussed in section 5.1.6 of Daniele and Zich (2014). We observe that (1.11.13) does not guarantee the absence of the singularities of $G^{-1}(\eta)$ located in $\mathrm{Im}[\eta] > 0$ as in the closed-form solution. These singularities are offending for the plus function $F_+(\eta)$ due to its regularity properties and, in fact, they do not appear in the exact closed-form solution. However, the contribution of these singularities in the representation (1.11.13) is due to the discretization procedure and their effect is reduced by increasing the numerical accuracy of the discretization of the integral.

1.12. Analytic continuation outside the integration line in the η-plane

Equation (1.11.13) holds for η located on the integration line $\lambda(y), \quad -\infty < y < \infty$ and by using an analytic continuation in its neighborhood.

While computing the field components of wedge scattering problems, we need a complete approximate representation of $F_+(\eta)$ in the complex plane η.

For example, the estimation of reflected and refracted plane waves, surface waves, lateral waves, leaky waves, diffraction coefficients, near field and so on requires the knowledge of the spectra, both in the proper and improper sheets of the complex plane η.

Let us interpret (1.11.5) in a more general form

$$\mathbf{F}_+(\eta) = -\int_{B_\theta} \mathbf{M}(\eta,\eta')\mathbf{F}_+(\eta')d\eta' + \mathbf{N}(\eta) \quad (1.12.1)$$

The integral $I(\eta) = \int_{B_\theta} \mathbf{M}(\eta,\eta')\mathbf{F}_+(\eta')d\eta'$ that appears in the representation defines a sectional holomorphic function (Gakhov 1990) for the presence of singularity lines in the kernel $\mathbf{M}(\eta,\eta')$. A singularity line is a line consisting of the values of the observation point η that make the kernel $\mathbf{M}(\eta,\eta')$ singular when η' travels on the integration line. This means that the analytic continuation of the integral $I(\eta)$ is valid starting from η located on the integration line until values of η do not cross a singularity line. Even though Plemelj equations (Gakhov 1990) allow the estimation of the jump of $I(\eta)$ when η crosses the singularity line, analytic continuation of (1.12.1) however, remains a complex task in the presence of singularity lines. In particular, since the function $\mathbf{F}_+(\eta)$ remains continuous when crossing the singularity line due to $\mathbf{M}(\eta,\eta')$, the problem consists of obtaining a suitable sectional holomorphic function $Z(\eta)$ such that $\mathbf{F}_+(\eta)$ remains continuous in representation (1.12.1).

In practice the analytic continuation of (1.12.1) beyond the first singularity line is

$$\mathbf{F}_+(\eta) = -\int_{B_\theta} \mathbf{M}(\eta,\eta')\mathbf{F}_+(\eta')d\eta' + \mathbf{Z}(\eta) + \mathbf{N}(\eta) \quad (1.12.2)$$

where the discontinuity of $I(\eta)$ is compensated by $Z(\eta)$.

In classical WH equations, considering the kernel $\dfrac{[G(\eta')-G(\eta)]}{\eta'-\eta}$ of (1.11.5), while both η and η' lie in the proper sheet, we have that $G(\eta) = G(\eta')$ when $\eta = \eta'$. However, when η lies on the improper sheet, $G(\eta)$ assumes values different from $G(\eta')$, particularly when $\eta = \eta'$ and the integration line is singular.

This means that when we wish to evaluate $\mathbf{F}_+(\eta)$ for η in the proper sheet, the representation (1.11.5) is valid. On the contrary, a process of analytic continuation is necessary for values of η located in the improper sheet. This process is better accomplished by introducing the complex plane w, which is defined in the following sections. In fact, in this plane, we can obtain simple recursive equations (difference equations) starting from the WH equations. Moreover, we anticipate that, for complex problems, singular lines of different kinds are also present when the observation point η is located in the proper sheet, due to the different mathematical properties of the kernel in the related FIE.

1.13. The complex plane w

The complex plane w is commonly used in diffraction problems studied with different techniques such as the Sommerfeld–Malyuzhinets technique where the Sommerfeld integral is a spectral representation in the w plane (see Chapter 2).

The mapping (1.13.1) also assumes an important role in wedge scattering problems studied with the WH technique:

$$\eta = -k\cos w \tag{1.13.1}$$

For example, the w plane can help handle and manipulate functions and functional properties that may be hard and obscure in the Laplace domain η. In the w plane, plus functions in η present important properties that allow the elimination of minus functions with an alternative route, with respect to what can be obtained from classical WH methodology and the Fredholm factorization method.

A detailed study of the mapping (1.13.1) has been presented in Daniele (2003b) and Daniele and Zich (2014).

In the following, we denote the quantities defined in the w plane with ^(hat) and the corresponding quantities defined in η with no label. For example, with respect to the spectral propagation multivalued function $\xi(\eta) = \sqrt{k^2 - \eta^2}$ (1.2.4), we have:

$$\hat{\xi}(w) = \xi(-k\cos w) = -k\sin w \tag{1.13.2}$$

with the assumption of the proper sheet for $\hat{\xi}(-\pi/2) = \xi(0) = k$, thus $\mathrm{Im}[\xi(\eta)] \le 0$.

As stated in section 1.2, multivalued functions of wedge scattering problems are defined in terms of $\xi(\eta)$; therefore, all the properties of the mapping (1.13.1) are related to the definitions of proper and improper sheets in $\xi(\eta)$.

The inversion of the mapping (1.13.1) introduces infinite replicates of w (Daniele 2003b; Daniele and Lombardi 2006; Daniele and Zich 2014):

$$w(\eta) = \arccos(-\frac{\eta}{k}) = -j\log(\frac{\eta + j\xi(\eta)}{k}) - \pi \tag{1.13.3}$$

No additional branch points with respect to those due to $\xi(\eta) = \sqrt{k^2 - \eta^2}$ arise from the logarithm. In fact, $\frac{\eta + j\xi(\eta)}{k}$ is never null as η is a finite point (note that $\eta + j\xi(\eta) = 0 \rightarrow \eta^2 = k^2 + \eta^2$).

To achieve a unique relation between a point η and its corresponding point w from (1.13.3), we need to uniquely define the properties of the log function: for each complex number z, we assume $-\pi < \arg[z] \leq +\pi$ such that $\log(z) = \log|z| + \arg(z)$.

According to (1.13.3), we need to define the value of $\xi(\eta) = \sqrt{k^2 - \eta^2}$ that depends on the selection of the branch lines that originate from $\eta = \pm k$. As in Daniele and Lombardi (2006) and Daniele and Zich (2014), in this section, we assume that the branch lines consist of the arcs of hyperbolas with $\mathrm{Im}[\xi(\eta)] = 0$ and the proper sheet with $\mathrm{Im}[\xi(\eta)] < 0$ (Figure 1.13.1).

With the assumptions on the branch of $\xi(\eta)$, the proper sheet and the principal value of $\log(z)$, we have the general relationship:

$$w(\eta) = \begin{cases} -j\log(\frac{\eta + j\xi(\eta)}{k}) - \pi, & \arg\frac{\eta + j\xi(\eta)}{k} > -\frac{\pi}{2} \\ -j\log(\frac{\eta + j\xi(\eta)}{k}) + \pi, & \arg\frac{\eta + j\xi(\eta)}{k} \leq -\frac{\pi}{2} \end{cases} \tag{1.13.4}$$

With these assumptions, the image of the proper and improper sheets of the η-plane in the w plane is presented in Figure 1.13.2, where the proper sheet is defined on the constraint $\mathrm{Im}[\xi(\eta)] < 0$. The image of the proper and improper sheets of the η-plane onto the w plane is shown in gray and blue colors, respectively, in Figure 1.13.2.

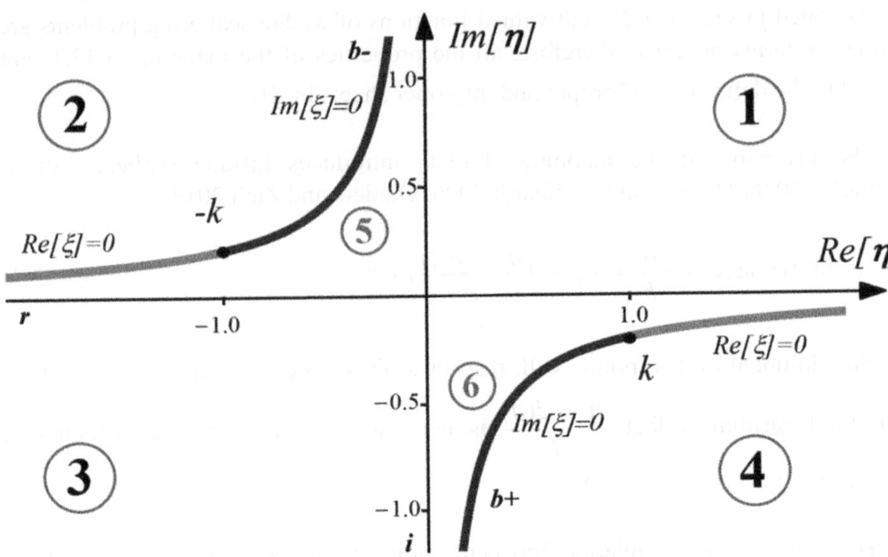

Figure 1.13.1. *Proper sheet of the η-plane with four quadrants 1, 2, 3 and 4 delimited by the images of the standard branch lines labeled with b+ and b- (we assume the branches with Im[ξ]=0 as standard). The proper sheet is defined on the constraint Im[ξ(η)]<0. Sub-regions 5 and 6 are respectively part of the second and fourth quadrants of the η-plane; however, they show different signs on the real part of ξ with respect to the rest of their quadrant. Here, k is equal to 1-j0.2. For a color version of the figure, see www.iste.co.uk/daniele/scattering1.zip*

Note that the proper and improper sheets of η are replicated periodically in the w plane.

Alternative options of branch lines are presented in section 1.2, which will be used in this book.

For example, when selecting the line $\mathrm{Re}[\xi(\eta)] = 0$ (see Figure 1.13.1) using certain technical computing software by default, the definition of the proper sheet is based on the constraint $\mathrm{Re}[\xi(\eta)] \geq 0$ instead of $\mathrm{Im}[\xi(\eta)] < 0$. With this assumption, the image of the proper and improper sheets of η in the w plane is different from that described previously and in Figure 1.13.2.

In this case, the proper-sheet is colored with light gray and dark blue.

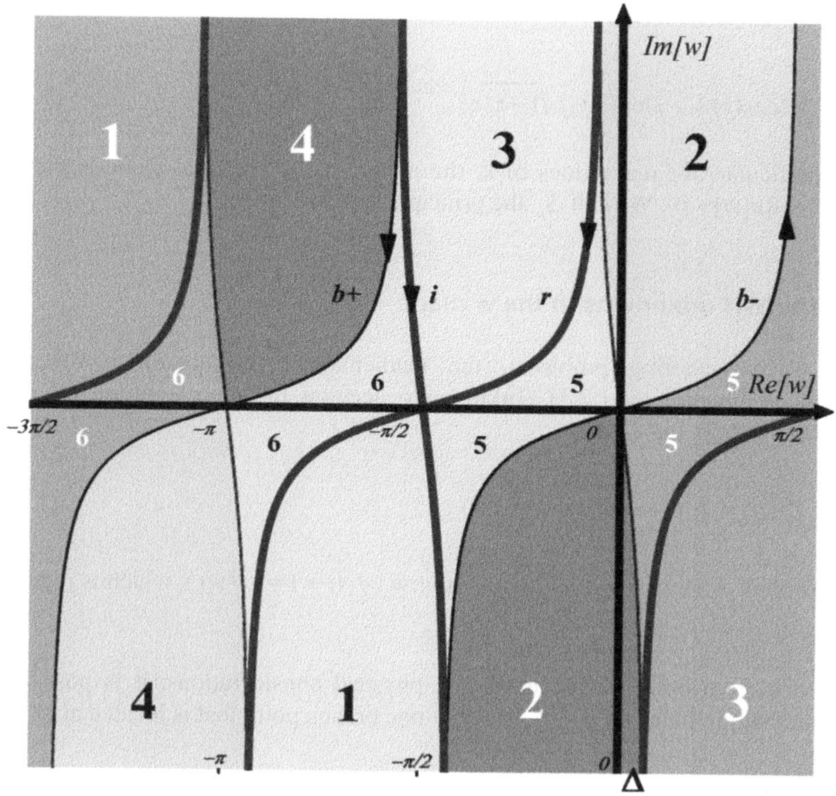

Figure 1.13.2. *Image of the proper and improper sheets of the η-plane onto the w plane is shown in gray and blue, respectively. For the proper sheet, the images of four quadrants 1, 2, 3 and 4 are delimited by the images of the standard branch lines b+ and b-. The r and i lines are the images of the real and imaginary axes of the η-plane. Sub-regions 5 and 6 are part of the second and fourth quadrants of the η-plane, respectively; however, they show different signs on the real part of ξ with respect to the rest of their quadrant. Here, k is equal to 1-j0.2. Δ is arctan(-Im[k]/Re[k]). The proper and improper sheets of the η-plane are replicated periodically in the w plane with the period 2π, For a color version of the figure, see www.iste.co.uk/daniele/scattering1.zip*

In general, when the point η and its corresponding w are located in the proper sheet, their relationship is simplified with respect to (1.13.4):

$$w(\eta) = -\arccos(-\frac{\eta}{k}) \tag{1.13.5}$$

assuming

$$\arccos(z) = -j\log(z + j\sqrt{1-z^2}\, e^{\frac{1}{2}j\arg(1-z^2)}) \tag{1.13.6}$$

In particular, for real values of k, the image of the principal plane is the strip $S_p : -\pi \leq \mathrm{Re}[w] \leq 0$. We call S_p the principal (proper) strip.

1.14. The WH unknowns in the w plane

In wave scattering problems, the mathematical formulation involves the multivalued function ξ (1.2.4). Taking into account the representation of a plus function $f_+(\eta)$ in the w plane

$$f_+(\eta) = f_+(-k\cos(w)) = \hat{f}_+(w) \tag{1.14.1}$$

we show that $\hat{f}_+(w)$ is an even function of w ($\hat{f}_+(-w) = \hat{f}_+(w)$), which is regular at w=0.

In fact, from both mathematical and physical considerations, it is possible to ascertain that the function $f_+(\eta)$ has only one branch point that is located at $\eta = k$.

From (1.13.2) and

$$\sqrt{k-\eta} = \sqrt{2k}\cos(w/2) \tag{1.14.2}$$

$$\eta = -k\cos(w) = k(1 - 2\cos^2(w/2)) \tag{1.14.3}$$

it follows that $\hat{f}_+(w)$ is a meromorphic function of w, i.e. $\hat{f}_+(w) = \dfrac{n(w)}{d(w)}$ with n(w) and d(w) being entire functions. It means that near $\eta = k$, i.e. w=$-\pi$, we have:

$$n(w) = a_0 + a_2 w^2 + a_4 w^4 + \ldots. \tag{1.14.4}$$

$$d(w) = b_0 + b_2 w^2 + b_4 w^4 + \ldots. \tag{1.14.5}$$

Since we are dealing with entire functions, (1.14.4) and (1.14.5) hold for every w.

It implies the evenness of $f_+(\eta) = n(w)/d(w)$ everywhere in the w plane.

Moreover, since the point w=0 corresponds to the regular point $\eta = -\tau_o$, $f_+(-k\cos(w))$ is regular at w=0.

To demonstrate properties on minus functions, we relate them to plus functions

$$f_-(\eta) = g_+(-\eta) \tag{1.14.6}$$

In the proper sheet defined in the w plane (1.13.2) we have

$$\xi(\eta) = \sqrt{k^2 - \eta^2} = \sqrt{k^2 - (-\eta)^2} = -k\sin(w) = \hat{\xi}(w) \tag{1.14.7}$$

where we have considered $w(-\eta) = -w(\eta) - \pi$, and $\xi(0) = k = \hat{\xi}(-\pi/2)$.

Therefore, by using the hat notation, it follows the properties for minus functions

$$g_+(-\eta) = \hat{g}_+(-w-\pi) \tag{1.14.8}$$

and considering the even symmetry of the plus function, we have

$$\hat{f}_-(w) = \hat{g}_+(-w-\pi) = \hat{g}_+(w+\pi) \tag{1.14.9}$$

This result also shows that a minus function does not change value in the w plane if w is substituted by $-2\pi - w$.

According to the definition of the principal strip S_p presented in section 1.13, once the WH problem in this strip is solved, we can extend the solution in the so-called fundamental strip $S_o : -\pi \le \mathrm{Re}[w] \le \pi$.

Moreover, we observe that, in the w plane, the WH equations become difference equations (see section 1.15). Using the methodology to solve difference equations, it is possible to obtain the solution for WH equations in the w plane that is alternative to the solutions obtained with the Fredholm factorization method reported previously and the classical solution in the Laplace domain when available.

However, while formulations of difference equations remain the fundamental tool to solve wedge scattering problems using the Sommerfeld–Malyuzhinets method (Budaev 1995; Babich *et al.* 2008), in this book, we apply these equations only to implement analytic continuations of approximate solutions obtained with the Fredholm factorization method (see section 1.15).

For example, we note that analytic continuations are easy to implement in the w plane (difference equations) with respect to analogous relations in the η-plane.

In particular, due to the presence of multivalued functions, we note that the same η value of the proper and improper sheets correspond to different values in the w plane (section 1.13). Therefore, the observation points can be easily tracked in the w plane with respect to the η-plane.

In summary, the Fredholm factorization method allows an efficient solution of classical WH equations in the proper sheet of the η-plane that can be easily and analytically extended in the improper sheet using difference equation formulations of the WH equations in the w plane.

Generalization will be implemented in the next chapters, where the Fredholm factorization method allows an efficient solution of generalized WH equations in a portion of the proper sheet of the η-plane.

1.15. The Fredholm factorization technique in the w plane

Equation (1.11.13) with the mapping (1.13.1) and an appropriate contour warping yields the approximate representation $\hat{F}_+^a(w)$ through an FIE of the second kind defined in the w plane. Alternative techniques derive Fredholm integral equations with some effort starting from difference equation formulations in the w plane (Budaev 1995; Babich *et al.* 2008; Lyalinov and Zhu 2013).

Let us consider the WH equation in which the source is a simple pole η_o:

$$G(\eta)F_+(\eta) = F_-(\eta) + \frac{F_o}{\eta - \eta_o} \qquad (1.15.1)$$

For simplicity, in the following we consider $\text{Im}[\eta_o] < 0$ in (1.15.1), thus from (1.9.6) we have the following Fredholm integral equation:

$$G(\eta)F_+(\eta) + \frac{1}{2\pi j}\int_{-\infty}^{\infty} \frac{[G(\eta') - G(\eta)]F_+(\eta')}{\eta' - \eta} d\eta' = \frac{F_o}{\eta - \eta_o} \qquad (1.15.2)$$

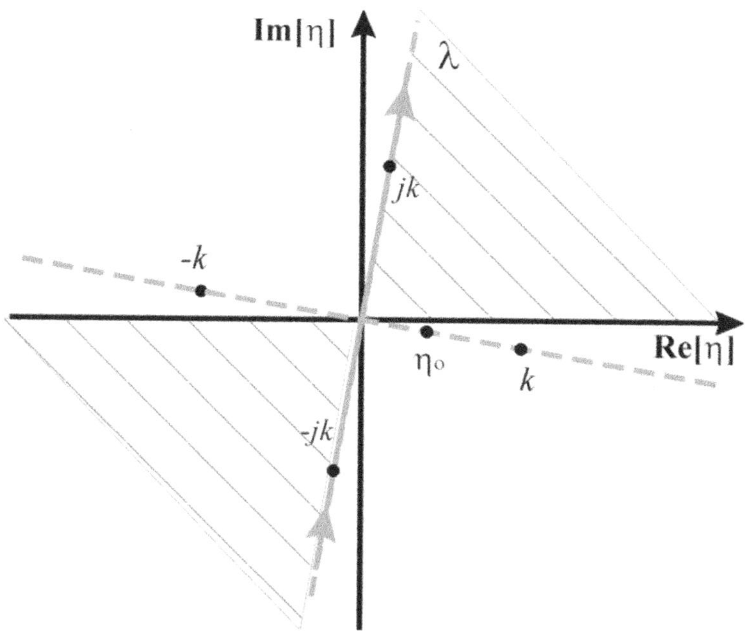

Figure 1.15.1. *Contour deformation of the integration line from the real axis of the η-plane to λ. For a color version of the figure, see www.iste.co.uk/daniele/scattering1.zip*

It is convenient to warp the contour path, as described in section 1.11. For example, we warp the real axis into the straight line λ shown in Figure 1.15.1: $\eta' = jky, \quad -\infty < y < \infty$. In wedge scattering problems, there are no singularities of $G(\eta)$ and $F_+(\eta)$ in the shaded region (i.e. in the first and third quadrants), which yield:

$$\frac{1}{2\pi j}\int_{-\infty}^{\infty}\frac{[G(\eta')-G(\eta)]F_+(\eta')}{\eta'-\eta}d\eta' = \frac{1}{2\pi j}\int_{\lambda}\frac{[G(\eta')-G(\eta)]F_+(\eta')}{\eta'-\eta}d\eta' \quad (1.15.3)$$

To obtain the FIE in the w plane, let us introduce the mappings (1.13.1) in the proper sheets:

$$\eta = -k\cos w, \qquad \eta' = -k\cos w' \quad (1.15.4)$$

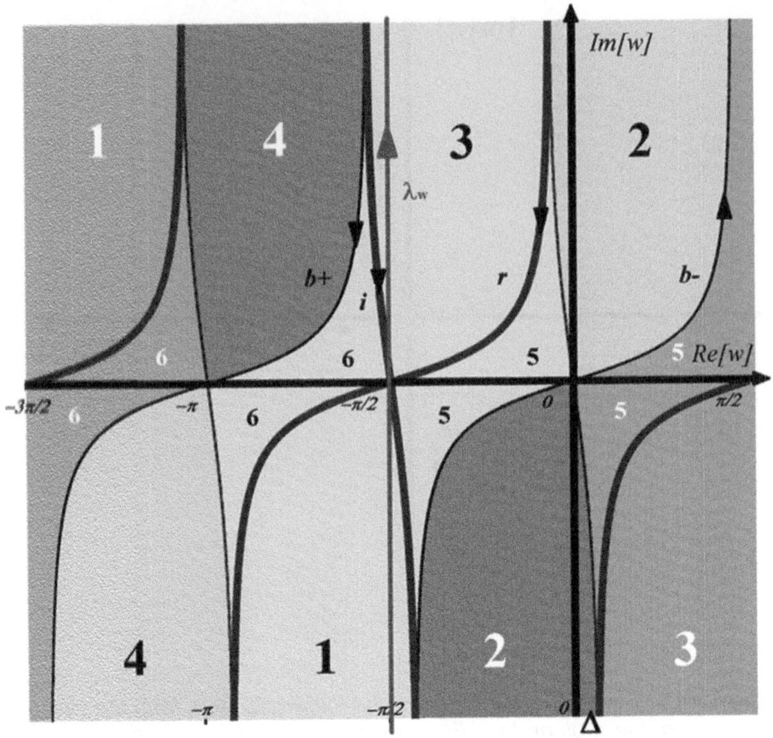

Figure 1.15.2. *Integration path λ in the w plane. For a color version of the figure, see www.iste.co.uk/daniele/scattering1.zip*

The pole η_o in wave scattering problems arises from a plane wave incident on a scatterer; therefore, we relate it to a corresponding singularity in the w plane and to an angle in the physical domain:

$$\eta_o = -k\cos w_o = -k\cos\varphi_o\,,\; w_o = -\varphi_o \tag{1.15.5}$$

where φ_o is the incidence angle. According to our assumptions, since k has a small imaginary part and $\mathrm{Im}[\eta_o] < 0$, we have $\pi/2 < \varphi_o < \pi$.

In the w plane, the Fredholm equation (1.15.2) becomes:

$$\hat{G}(w)\hat{F}_+(w) - \frac{1}{2\pi j}\int_{\lambda_w} \frac{[\hat{G}(w') - \hat{G}(w)]\hat{F}_+(w')}{-k\cos w' + k\cos w} k\sin w'\,dw' = -\frac{F_o}{k\cos w - k\cos w_o} \tag{1.15.6}$$

where λ_w is the image in the w plane of the line λ of the η plane paying attention on the orientation of the curves. Taking into account (1.15.4), we obtain that λ_w is a vertical line defined by the parameterization $w' = -\pi/2 + ju$ with $-\infty < u < +\infty$. It yields $dw' = jdu$, $\cos w' = j\sinh u$, $\sin w' = -\cosh u$ and

$$\hat{G}(w') = H(u), \qquad \hat{F}_+(w) = Y(u) \tag{1.15.7}$$

Thus, (1.15.6) assumes the form:

$$\hat{G}(w)\hat{F}_+(w) + \frac{1}{2\pi j}\int_{-\infty}^{+\infty} \frac{\left[H(u) - \hat{G}(w)\right] Y(u)}{-j\sinh u + \cos w}\cosh u \; jdu = -\frac{F_o}{k\cos w - k\cos w_o} \tag{1.15.8}$$

where $w \in \lambda_w$ and it provides the representation:

$$\hat{F}_+(w) = -\frac{1}{2\pi j}\hat{G}^{-1}(w)\int_{-\infty}^{+\infty} \frac{\left[H(u) - \hat{G}(w)\right] Y(u)}{-j\sinh u + \cos w}\cosh u \; jdu - \frac{\hat{G}^{-1}(w)F_o}{k\cos w - k\cos w_o} \tag{1.15.9}$$

Let us now deduce the equations valid for the samples $Y(u)$ (1.15.7). Putting $w = -\pi/2 + jt$ in (1.15.9) yields:

$$H(t)Y(t) + \frac{1}{2\pi j}\int_{-\infty}^{+\infty} m(t,u)Y(u)du = -\frac{F_o}{k(j\sinh t - \cos w_o)}, \quad t \in \mathbb{R} \tag{1.15.10}$$

where:

$$m(t,u) = \begin{cases} \lim\limits_{u\to t} \dfrac{\left[H(u) - H(t)\right]}{-\sinh u + \sinh t}\cosh u, & u = t \\ \dfrac{\left[H(u) - H(t)\right]}{-\sinh u + \sinh t}\cosh u, & u \neq t \end{cases} \tag{1.15.11}$$

The sampled version of (1.15.10) is.

$$H(hr)Y_a(hr) + \frac{h}{2\pi j}\sum_{i=-A/h}^{A/h} m(hr,hi)Y_a(hi) = -\frac{F_o}{k(j\sinh hr - \cos w_o)} \tag{1.15.12}$$

with $r \in [-A/h, A/h]$ (r,i integer).

Note that the observation point and the integration point are taken on the same line λ_w in this discretized form of the Fredholm integral equation.

In (1.15.12), the quantities h and $\pm A$ represent, respectively, the discretization step and the endpoints of the truncated integration line for the approximation.

Equation (1.15.12) constitutes a linear system of order 2A+1 similar to that described in section 1.11. The solution of the linear system provides approximate samples of $Y_a(hr)$. It yields the approximate solution

$$Y_a(t) = -\frac{H^{-1}(t)h}{2\pi j}\sum_{i=-A/h}^{A/h} m(t,hi)Y_a(hi) - \frac{H^{-1}(t)F_o}{k(j\sinh t - \cos w_o)} \qquad (1.15.13)$$

which allows us to represent an approximate version of the plus unknown $\hat{F}_+(w)$ as $\hat{F}_{a+}(w) = Y_a(u)$:

$$\hat{F}_{a+}(w) = -\frac{G^{-1}(w)h}{2\pi j}\sum_{i=-A/h}^{A/h} m(-j(w+\frac{\pi}{2}),hi)Y_a(hi) - \frac{G^{-1}(w)F_o}{k(\cos w - \cos w_o)} \qquad (1.15.14)$$

1.16. Analytic continuation in the w plane

The solution of the Fredholm integral equation directly obtained in the w plane, as in section 1.15, or obtained in the η-plane, as in section 1.11, and mapped into the w plane via (1.13.1) needs analytic continuation to obtain values of the solution in both complete proper and improper sheets. Moreover, the approximate representation $\hat{F}_+^a(w)$ holds for each value of w until w does not cross the singularity lines of the kernel. In the case of generalized WH equations, the restriction may also include singularity lines in the proper sheet.

Rephrasing the analysis in the w plane of section 1.12, in classical WH equations, considering the kernel $\frac{[G(\eta')-G(\eta)]}{\eta'-\eta}$ and the related FIE in the w plane, singularity lines are the vertical lines $w = \pm w' + 2n\pi$, $w' \in \lambda_w$, where n is a non-null integer. We recall that the integral representation is a sectional analytic representation in w due to the integral term (Gakhov 1990). As soon as w crosses one of these lines, the integral produces a discontinuity.

Since the approximate solution holds in the proper sheet and the discontinuity of singularity lines must be compensated, we need to enforce the analytic continuation of $\hat{F}_+(w)$ in the whole complex plane w, i.e. $\hat{F}_+(w)$ must be analytic and continuous except for physical singularities.

In sections 1.11 and 1.12, we outlined the analytic continuation procedure in the η-plane, and here we present the same procedure in the w plane using an alternative and simple method, based on difference equations obtained from the properties presented in sections 1.13 and 1.14 for the WH unknowns.

Difference equations can be easily obtained from WH equations. Taking into account the classical WH equation:

$$G(\eta)F_+(\eta) = X_+(-\eta) \tag{1.16.1}$$

where $F_-(\eta) = X_+(-\eta)$, in the w plane, we have

$$\hat{G}(w)\hat{F}_+(w) = \hat{X}_+(w+\pi) \tag{1.16.2}$$

Or alternatively

$$\hat{G}(w-\pi)\hat{F}_+(w-\pi) = \hat{X}_+(w) \tag{1.16.3}$$

We can eliminate $\hat{X}_+(w)$ by taking into account the properties of WH plus unknowns in the w plane, i.e. its evenness $\hat{X}_+(-w) = \hat{X}_+(w)$. It yields:

$$\hat{G}(w-\pi)\hat{F}_+(w-\pi) = \hat{G}(-w-\pi)\hat{F}_+(-w-\pi) = \hat{G}(-w-\pi)\hat{F}_+(w+\pi) \tag{1.16.4}$$

or if we substitute w with $w-\pi$

$$\hat{F}_+(w) = \left[\hat{G}(-w)\right]^{-1} \hat{G}(w-2\pi)\hat{F}_+(w-2\pi) \tag{1.16.5}$$

Equation (1.16.5) presents a recursive relation that is particularly useful for analytic continuation if the approximated estimation of $\hat{F}_+(w)$ is obtained in a strip of 2π wideness for real values of w.

We note that if $\hat{F}_{a+}(w)$ is valid in the principal strip S_p (section 1.13), the evenness of $\hat{F}_+(w)$ allows us to obtain an approximated value on the fundamental strip $S_o : -\pi \le \mathrm{Re}[w] \le \pi$ (section 1.14).

Finally, the application of the recursive relation, along with the symmetry of $\hat{F}_+(w)$, allows us to obtain a representation in the whole complex plane w (proper/improper sheets and their replica).

Generalized versions of (1.16.5) for generalized WH equations will be provided in the next chapters.

We note that one of the principal reasons for introducing the w plane is the possibility of reducing the wedge problems to the knowledge of the spectra in the principal strip S_p.

1.17. The Fredholm factorization technique to factorize the kernel

An important application of the Fredholm factorization method is the approximate estimation of the factorization of the matrix WH kernel $G(\eta)$ of order n (Daniele and Lombardi 2007):

$$G(\eta) = G_-(\eta)G_+(\eta) \tag{1.17.1}$$

Let us introduce the n homogeneous matrix WH equation problems of order n (Vekua 1967):

$$G(\eta)U_{i+}(\eta) = U_{i-}(\eta) \qquad i=1,2,...n \tag{1.17.2}$$

which can be rewritten in the compact form

$$G(\eta)U_+(\eta) = U_-(\eta) \tag{1.17.3}$$

where $U_+(\eta)$ and $U_-(\eta)$ are matrices of order n:

$$U_+(\eta) = |U_{1+}(\eta), U_{2+}(\eta), ...U_{n+}(\eta)| \tag{1.17.4}$$

$$U_-(\eta) = |U_{1-}(\eta), U_{2-}(\eta), ...U_{n-}(\eta)| \tag{1.17.5}$$

By defining

$$U_+(\eta) = |U_{1+}(\eta), U_{2+}(\eta), ...U_{n+}(\eta)| = G_+^{-1}(\eta) \tag{1.17.6}$$

$$U_-(\eta) = |U_{1-}(\eta), U_{2-}(\eta), ...U_{n-}(\eta)| = G_-(\eta) \tag{1.17.7}$$

$G_+^{-1}(\eta)$ and $G_-(\eta)$ are invertible plus and minus matrices when the vectors $U_{i+}(\eta)$ and $U_{i-}(\eta)$ are independent.

Since

$$G(\eta)G_+^{-1}(\eta) = G_-(\eta) \tag{1.17.8}$$

$$G(\eta) = G_-(\eta)G_+(\eta) \tag{1.17.9}$$

we have

$$G_+(\eta) = |U_{1+}(\eta), U_{2+}(\eta), ...U_{n+}(\eta)|^{-1} \tag{1.17.10}$$

$$G_-(\eta) = |U_{1-}(\eta), U_{2-}(\eta), ...U_{n-}(\eta)| \tag{1.17.11}$$

$G_+(\eta)$ and $G_-(\eta)$ are necessarily the two factorized matrix functions of the kernel $G(\eta)$.

In order to estimate $U_{i+}(\eta)$ and $U_{i-}(\eta)$, we apply the Fredholm factorization method.

Since $U_{i+}(\eta)$ and $U_{i-}(\eta)$ do not vanish at infinity they are thus not Laplace transforms. For this reason we define auxiliary Laplace transform unknowns:

$$X_{i+}(\eta) = \frac{U_{i+}(\eta)}{\eta - \eta_p} \tag{1.17.12}$$

with $\text{Im}[\eta_p] < 0$ such that $\eta_p \notin Null[G(\eta)] \cup Null[G^{-1}(\eta)]$.

Note that if $\eta_q \in Null[G(\eta)]$, then $U_{i-}(\eta_q) = 0$ and if $\eta_q \in Null[G^{-1}(\eta)]$, then $U_{i+}(\eta_q) = 0$.

$$Null[G(\eta)] = \{\eta : \det[G(\eta)] = 0\} \tag{1.17.13}$$

$$Null[G^{-1}(\eta)] = \{\eta : \det[G^{-1}(\eta)] = 0\} \tag{1.17.14}$$

By substituting (1.17.12) into (1.17.2), we have

$$G(\eta)X_{i+}(\eta) = \frac{U_{i-}(\eta)}{\eta - \eta_p} \tag{1.17.15}$$

and, after some mathematical manipulations we have

$$G(\eta)X_{i+}(\eta) = \frac{U_{i-}(\eta) - U_{i-}(\eta_p)}{\eta - \eta_p} + \frac{U_{i-}(\eta_p)}{\eta - \eta_p} \tag{1.17.16}$$

By taking $U_{i-}(\eta_p) = R_i$, where R_i is the canonical basis of R^n, we have n independent problems, which yield n independent solutions related to η_p and

$$U_{i+}(\eta) = X_{i+}^p(\eta)(\eta - \eta_p) \tag{1.17.17}$$

If closed-form factorization is available, we have from (1.5.13)

$$X_{i+}^p(\eta) = \frac{G_+^{-1}(\eta)G_-^{-1}(\eta_p)}{\eta - \eta_p} R_i \tag{1.17.18}$$

otherwise, by implementing the Fredholm factorization method from (1.17.16), we obtain

$$X_{i+}^p(\eta) = -\frac{G^{-1}(\eta)}{2\pi j} \int_{-\infty}^{\infty} \frac{[G(\eta') - G(\eta)] X_{i+}^p(\eta')}{\eta' - \eta} d\eta' + \frac{G^{-1}(\eta) R_i}{\eta - \eta_p} \tag{1.17.19}$$

by eliminating the minus function $\dfrac{U_{i-}(\eta) - U_{i-}(\eta_p)}{\eta - \eta_p}$.

Note that this procedure is interesting only when closed-form factorization is not available.

With the knowledge of $X_{i+}^p(\eta)$, we obtain an approximation of $G_+(\eta)$ using (1.17.17) and (1.17.10).

Let us label the factorized matrix functions related to η_p $G_+(\eta, \eta_p)$ and $G_-(\eta, \eta_p)$.

We prove that two selections of η_p yield factorized matrix functions that differ from a constant matrix:

$$G_+(\eta,\eta_2) = KG_+(\eta,\eta_1) \tag{1.17.20}$$

with

$$K = \left|G_-^{-1}(\eta_2)R_1, G_-^{-1}(\eta_2)R_2, ...G_-^{-1}(\eta_2)R_n\right|^{-1} \left|G_-^{-1}(\eta_1)R_1, G_-^{-1}(\eta_1)R_2, ...G_-^{-1}(\eta_1)R_n\right| \tag{1.17.21}$$

K is obtained using (1.17.18), (1.17.12) and (1.17.10). In fact, by substituting (1.17.18) into (1.17.12), we have

$$U_{i+}^p(\eta)[G_-^{-1}(\eta_p)R_i]^{-1} = G_+^{-1}(\eta) \tag{1.17.22}$$

and after equating (1.17.22) for p=1,2, we have from (1.17.10) the definition of K.

1.18. Some examples of the Fredholm factorization method

In this section, we validate the Fredholm factorization method through examples that are drawn from pure mathematics and applied physics.

For example, we show that the approximate factorizations, obtained from the numerical solution of Fredholm integral equations as described in the previous sections, are good approximations of closed-form factorizations when available. We recall that the spectrum of WH kernels is related to the nature of structural singularities (branch points and/or poles).

Example 1: solution of a scalar WH equation

The first example is derived from the solution of the diffraction/scattering problem of a truncated planar PEC waveguide of thickness d immersed in a free space medium with propagation constant k. The formulation of the problem in terms of WH is classical and scalar

$$g(\eta)f_+(\eta) = f_-(\eta) \tag{1.18.1}$$

where the kernel is

$$g(\eta)=\frac{1}{\sqrt{k^2-\eta^2}}\frac{e^{j\sqrt{k^2-\eta^2}d}}{\cos[\sqrt{k^2-\eta^2}d]} \tag{1.18.2}$$

We note that the kernel contains both poles (1.18.3) and branches with the branch point at $\pm k$.

$$\eta_n=\pm\sqrt{k^2-\left(\frac{(n-1/2)\pi}{d}\right)^2},\quad n\in\mathbf{N}_0 \tag{1.18.3}$$

In order to apply the Fredholm factorization and the analytical factorization, we can first pre-factorize

$$\sqrt{k^2-\eta^2}=\sqrt{k-\eta}\sqrt{k+\eta} \tag{1.18.4}$$

which appear at the denominator of (1.18.2). We then need to factorize

$$G(\eta)=\frac{e^{j\sqrt{k^2-\eta^2}d}}{\cos[\sqrt{k^2-\eta^2}d]}=G_-(\eta)G_+(\eta) \tag{1.18.5}$$

where $g(\eta)=G(\eta)/\sqrt{k^2-\eta^2}$.

The WH problem can be reformulated with the following equation:

$$G(\eta)F_+(\eta)=F_-(\eta) \tag{1.18.6}$$

where $G(\eta)$ is a meromorphic function and $F_+(\eta)=f_+(\eta)/\sqrt{k-\eta}$ and $F_-(\eta)=f_-(\eta)\sqrt{k+\eta}$. To perform the analytical factorization of $G(\eta)$, we use the Mittag–Leffler (ML) expansion (Daniele and Zich 2014; Daniele and Lombardi 2007) for the singularities (1.18.3).

By assuming an illumination via the fundamental mode, we obtain a non-standard minus function with pole η_1 (1.18.3).

The solution is then obtained through (1.5.13) and (1.11.13) in terms of $F_+(\eta),F_-(\eta)$.

The application of the Fredholm factorization method (see sections 1.9–1.11) allows us to obtain an approximate estimation of $F_+(\eta)$ by eliminating $F_-(\eta)$.

Focusing the attention on a specific problem, we select k=1-j0.01 and d=0.55λ where the first mode of the parallel PEC waveguide is the only one that propagates.

In this example, we first apply the ML procedure by considering the first 200 poles in $G_{\pm}(\eta)$ and then the Fredoholm factorization method with the contour warping (1.11.4) with θ=π/4 and discretization parameters A=100 and h=0.4 (1.11.6). For the sake of simplicity, we assume that the non-standard part in (1.18.6) is normalized to 1, i.e.

$$F_o(\eta) = \frac{1}{\eta - \eta_1} \tag{1.18.7}$$

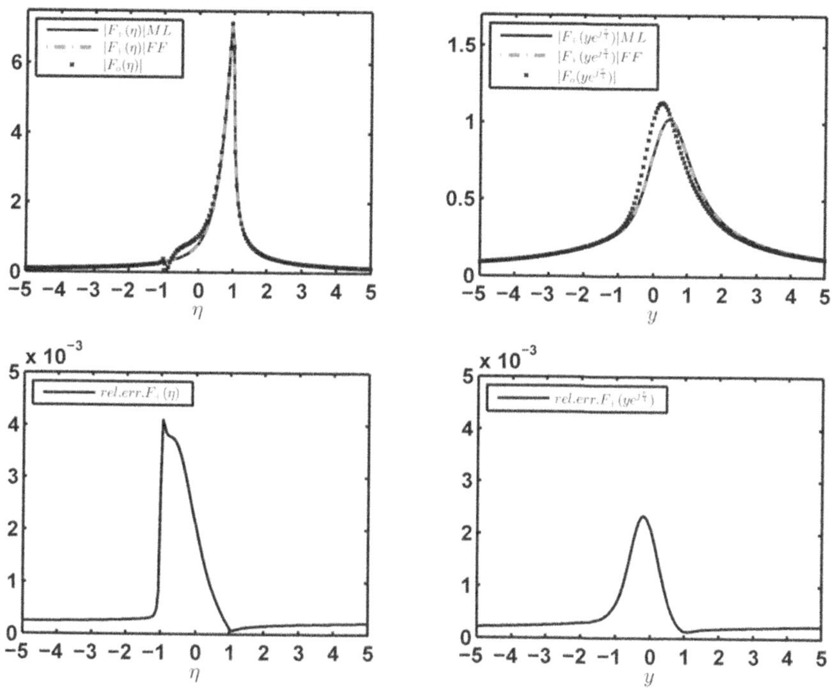

Figure 1.18.1. *Top left: the absolute value of $F_+(\eta)$ obtained with the Mittag–Leffler (ML) procedure, the Fredholm factorization (FF) method and the plot of its source term $F_o(\eta)$. Top right: the absolute value of $F_+(\eta)$ along the integration line λ, i.e. $F_+(e^{j\frac{\pi}{4}}y)$ obtained with the ML procedure, the FF method and the plot of its source term $F_o(e^{j\frac{\pi}{4}}y)$. Bottom: comparison between the two approximate solutions along the real axis of η and the contour λ*

In Figure 1.18.1, we present the comparison between the two approximate solutions. The numerical results demonstrate perfect agreement, i.e. relative errors are less than 1%. For this problem, we have selected A and h, making a compromise between computational cost and accuracy that can be further improved by diminishing h and increasing A, paying attention not to compromise the conditioning of the linear system. The values of A and h can be decided *a priori* by investigating the behavior of the kernel (1.11.7) of the associated Fredholm integral equation.

Example 2: rational matrix kernel WH factorization and solution of a matrix WH equation

In this example, we first present the application of the Fredholm factorization method to factorize the matrix WH kernel:

$$G(\eta) = \begin{pmatrix} 1 & jq\dfrac{\alpha^2+\eta^2}{\beta^2+\eta^2} \\ jq & 1 \end{pmatrix} = G_-(\eta)G_+(\eta) \tag{1.18.8}$$

This problem is directly related to the solution of the matrix WH equation (1.4.1) according to the theory presented in section 1.17.

In order to compare the Fredholm factorization, we perform the factorization in closed form using the method based on singularity matching reported in Daniele and Lombardi (2007) and Daniele and Zich (2014). It yields the closed-form factorization

$$G_+(\eta) = \begin{vmatrix} 0 & \dfrac{(-\beta\sqrt{1+q^2}+\sqrt{\beta^2+\alpha^2 q^2})(\sqrt{\beta^2+\alpha^2 q^2}-j\sqrt{1+q^2}\,\eta)}{q\left(\beta^2+\alpha^2 q^2-j\sqrt{1+q^2}\,\eta\right)} \\ \beta q-\dfrac{q\sqrt{\beta^2+\alpha^2 q^2}}{\sqrt{1+q^2}} & -\dfrac{\left(\beta q-\dfrac{q\sqrt{\beta^2+\alpha^2 q^2}}{\sqrt{1+q^2}}\right)(jB+\eta)}{q(\beta-j\eta)} \end{vmatrix} \tag{1.18.9}$$

and $G_-(\eta) = G(\eta)G_+^{-1}(\eta)$.

In order to apply the procedure presented in section 1.17, we need to analyze the null spaces of $G(\eta)$ and $G^{-1}(\eta)$ that are

$$Null[G(\eta)] = \{\eta : \det[G(\eta)] = 0\} = \{-\frac{\sqrt{-\beta^2 - \alpha^2 q^2}}{\sqrt{1+q^2}}, \frac{\sqrt{-\beta^2 - \alpha^2 q^2}}{\sqrt{1+q^2}}\} \quad (1.18.10)$$

$$Null[G^{-1}(\eta)] = \{\eta : \det[G^{-1}(\eta)] = 0\} = \{-j\beta, j\beta\} \quad (1.18.11)$$

We note that the singular points lie on the imaginary axis for real α, β, q.

Let us now focus on a specific problem: we select α=2, β=1 and q=0.75.

As a first step, we arbitrarily need to select $\eta_p \notin Null[G(\eta)] \cup Null[G^{-1}(\eta)]$, with $\mathrm{Im}[\eta_p] < 0$ yielding a non-standard minus behavior in the problem. According to the parameters of the problem, we select $\eta_p = 1 - j0.1$ since $Null[G(\eta)] \cup Null[G^{-1}(\eta)] = \{\pm j, \pm j1.442\}$. Due to the location of singularities in this problem, we do not warp the original contour (real axis) (1.11.4) and we select the discretization parameters of Fredholm factorization A=100, h=0.1 (1.11.6).

We note that the selection of η_p, A and h is a compromise between computational cost and accuracy that can be decided *a priori* by investigating the behavior of the kernel (1.11.7) of the associated Fredholm integral equation.

In Figure 1.18.2, we present the behavior of the factorized kernel $G_+(\eta)$ that is defined up to a multiplicative constant matrix K. According to (1.17.20), we can compare the closed form plus kernel $G_+(\eta)$ with that obtained using the Fredholm factorization method $G_+(\eta,\eta_p)$ by imposing the equality in a spectral point η (e.g. η=0)

$$KG_+(\eta = 0) = G_+(\eta = 0, \eta_p) \quad (1.18.12)$$

It yields

$$K \simeq \begin{vmatrix} 0.445 + j2.759 & -3.015 \\ 0 & -j2.261 \end{vmatrix} \quad (1.18.13)$$

$$KG_+(\eta) = \begin{vmatrix} 1 & \dfrac{(0.328+j2.034)(1.803-\dfrac{5j\eta}{4})}{1-\mathrm{i}\alpha} - \dfrac{4(j+\eta)}{3(1-j\eta)} \\ \dfrac{3j}{4} & -\dfrac{j(j+\alpha)}{1-j\alpha} \end{vmatrix} \tag{1.18.14}$$

Since three out of four components of $KG_+(\eta)$ are constant, in Figure 1.18.2 we present the comparison for the first row, i.e. the comparison between $(KG_+(\eta))_{1,s}$ and $(G_+(\eta,\eta_p))_{1,s}$ with s=1,2.

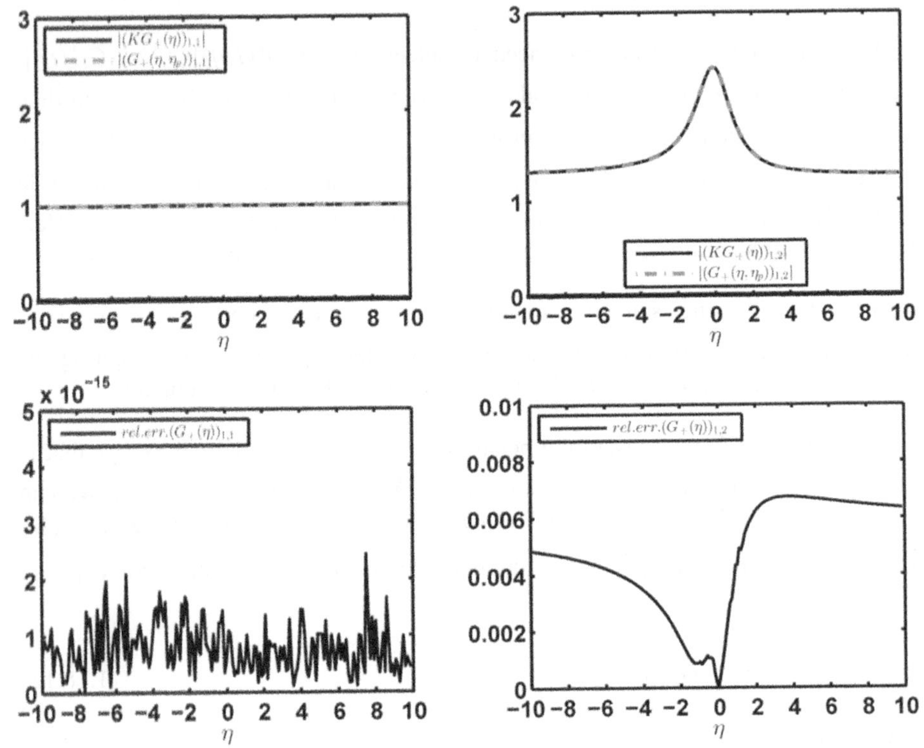

Figure 1.18.2. *Top: the absolute value of the component (i,j) of the plus factorized kernel (left (1,1), right (1,2)) obtained with the closed-form expression (1.18.14) and Fredholm factorization. Bottom: relative error between the component (i,j) of the plus factorized kernel (left (1,1), right (1,2)) obtained with the closed-form expression and Fredholm factorization*

In particular, the component (1,1) highlights the capability of Fredholm factorization to capture the constant behavior. The numerical results demonstrate perfect agreement, i.e. relative errors are also less than 1% for component (1,2).

In the second part of this example, we compare the solution of the matrix WH equation (1.4.1) in terms of $F_+(\eta)$ obtained with three different methods: 1) the closed-form solution using (1.5.13) and (1.18.9), 2) the approximate solution using (1.5.13) and the Fredholm factorized kernel $G_+(\eta,\eta_p)$, as discussed in the previous part of this example, and 3) the direct solution using the Fredholm factorization method, as described in section 1.11.

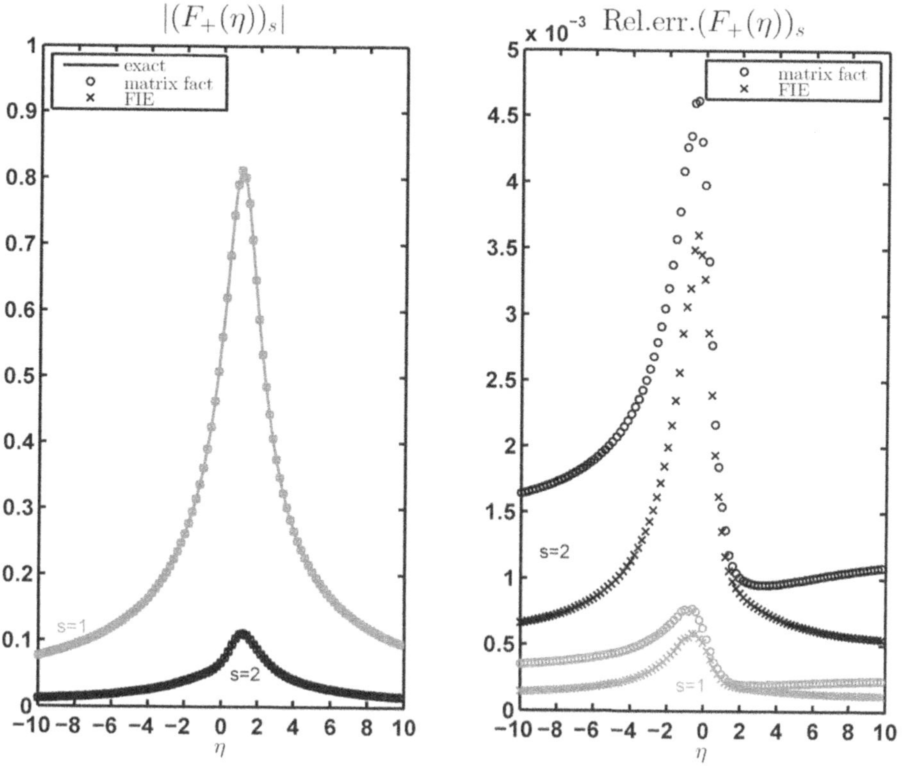

Figure 1.18.3. *Left: the absolute value of the component s (s=1 gray symbols, s=2 black symbols) of the plus vector function $F_+(\eta)$ obtained with the exact factorization (solid lines), the Fredholm factorized kernel (circles) and the direct solution of the FIE (crosses). Right: the relative error for the components (s=1 gray symbols, s=2 black symbols) of the solution $F_+(\eta)$ obtained by the Fredholm factorized kernel (circles) and the direct solution of the FIE (crosses) with respect to the exact solution. For a color version of the figure, see www.iste.co.uk/daniele/scattering1.zip*

In this example, we assume a non-standard minus source in (1.4.1):

$$F_o(\eta) = \frac{1}{\eta - \eta_o} \begin{vmatrix} j \\ -0.5 \end{vmatrix} \tag{1.18.15}$$

with $\eta_o = 1 - j$ and the same parameters α=2, β=1 and q=0.75 being selected previously.

The absolute values of each component of the plus vector function $F_+(\eta)$ are presented in Figure 1.18.3, along with the computation of relative errors. In the numerical results, we assume A=100, h=0.1 for the discretization of Fredholm integral equations and $\eta_p = 1 - j0.1$ for the Fredholm factorized kernel $G_+(\eta, \eta_p)$.

From Figure 1.18.3, we note that the numerical results demonstrate perfect agreement with the exact solution, i.e. relative errors are less than 1% for both components s=1,2.

For example, the direct solution performs better than the indirect solution obtained after the estimation of the Fredholm factorized kernel. This behavior is probably due to the propagation of errors. However, the approximate factorization of kernels is a useful tool to obtain the solution of the WH problem in a traditional form (1.5.13) for an arbitrary source, in particular, when a closed-form factorization of the kernel is not available.

Appendix 1.A: effect of using j and i as imaginary units in Fourier transforms

Throughout this book, we only consider time harmonic fields with a time dependence specified by the factor $e^{j\omega t}$ (electrical engineering notations), which is omitted, where the imaginary unit is indicated by j. Applied mathematics often uses a different notation $e^{-i\omega t}$, where the imaginary unit is indicated by i. This means that, in the natural domain, the change $j \Rightarrow -i$ transforms engineering notations in applied mathematics notations (and vice versa).

In fact, regarding, for example, the Fourier transforms, the following definitions are the most frequently used in the literature:

$$F_e(\eta) = \int_{-\infty}^{\infty} f_e(x) e^{j\eta x} dx, \quad F_a(\eta) = \int_{-\infty}^{\infty} f_a(x) e^{i\eta x} dx, \tag{1.A.1}$$

where the subscript e denotes engineering and the subscript a denotes applied mathematics. Consequently, in the spectral domain, we have:

$$F_a(\eta) = F_e(-\eta) \text{ and j substituted by –i (and vice versa)} \tag{1.A.2}$$

For example, let us consider in the natural domain the x-progressive propagation factor that is defined in electrical engineering notations by:

$$f_e(x) = e^{-jkx} \tag{1.A.3}$$

with the propagation constant k defined by:

$$k = \beta - j\alpha \quad , \quad \alpha \geq 0 \tag{1.A.4}$$

The same propagation factor in applied mathematics notations is written as:

$$f_a(x) = e^{ik_a x} \tag{1.A.5}$$

with $k_a = \beta + i\alpha$

In the Laplace domain, on the real axis, we have:

$$F_e(\eta) = \int_{-\infty}^{\infty} e^{-jkx} e^{j\eta x} dx = \frac{j}{\eta - k} \tag{1.A.6}$$

which in applied mathematics notations is written as:

$$F_e(\eta) = \frac{j}{\eta - k} \Rightarrow F_a(\eta) = \frac{-i}{-\eta - k_a} = \frac{i}{\eta + k_a} \tag{1.A.7}$$

Analytic continuations define the previous functions in the whole complex plane η. This means that the Laplace transforms are defined for any value of η by:

$$F_e(\eta) = \frac{j}{\eta - k}, \quad F_a(\eta) = \frac{i}{\eta + k_a} \tag{1.A.8}$$

In this book, we define plus $F_+(\eta)$ and minus $F_-(\eta)$ (see section 1.2). Note that a plus (or minus) function in electrical engineering notations is also a plus (or minus) function in applied mathematics notations. The only difference is given by the position of the singularities. For example, $F_e(\eta)$ and $F_a(\eta)$ defined in (1.A.8) are plus functions in both the "engineering" and "applied mathematics" notations. In fact, $F_e(\eta)=\dfrac{j}{\eta-k}$ has a singularity in $\eta=k=\beta-j\alpha$ and $F_a(\eta)=\dfrac{i}{\eta+k_a}$ in $\eta=-k_a=-\beta-j\alpha$. The properties described in this Appendix are extendable to definitions based on Laplace transforms.

Appendix 1.B: compactness of the matrix kernel $\dfrac{G^{-1}(\alpha)G(u)-\mathbf{1}}{u-\eta}$

With the conditions that $G(\eta)$ and $G^{-1}(\eta)$ exist and are bounded on the real axis as $\alpha\to\pm\infty$, in the following, we will state that the kernel $\dfrac{G^{-1}(\eta)G(u)-\mathbf{1}}{u-\eta}$ of the integral equations described in (1.9.6) and used in this book is compact.

First, the kernel is rewritten as:

$$\frac{G^{-1}(\eta)G(u)-\mathbf{1}}{u-\eta}=G^{-1}(\eta)\frac{G(u)-G(\eta)}{u-\eta} \tag{1.B.1}$$

Since $G(\eta)$ and $G^{-1}(\eta)$ are bounded, an important property of the functional analysis (Krasnov *et al.* 1983) reduces the proof of compactness of (1.B.1) to that of $\dfrac{G(u)-G(\eta)}{u-\eta}$.

The discussion of this kernel is considered in Gohberg and Krupnik (1992). In that study, the compactness of this kernel has been studied in a general context. Concerning our applications, it is sufficient to show that (Krasnov *et al.* 1983):

$$\int_{-\infty}^{\infty}\int_{-\infty}^{\infty}\left|\frac{[G_{rs}(u)-G_{rs}(\eta)]}{u-\eta}\right|^2 du\, d\eta<\infty \tag{1.B.2}$$

where $G_{rs}(u)$ are components of the matrix $G(u)$. Taking into account that the difficulty in proving the properties of (1.9.6) arises from the presence of an infinite integration interval, we only consider the contributions of

$$\int_M^\infty \int_M^\infty \left| \frac{[G_{rs}(u) - G_{rs}(\eta)]}{u-\eta} \right|^2 du\, d\eta \text{ and } \int_{-M}^M \int_M^\infty \left| \frac{[G_{rs}(u) - G_{rs}(\eta)]}{u-\eta} \right|^2 du\, d\eta \quad (1.B.3)$$

where M is a sufficiently large number but bounded.

Let us assume that in the interval $M \le x < \infty$, $G_{rs}(u)$ is represented by the convergent series

$$G_{rs}(u) = A_{rs}^0 + \frac{A_{rs}^1}{u^\nu u} + \frac{A_{rs}^2}{u^\nu u^2} + \ldots = A_{rs}^0 + \frac{1}{u^\nu} \sum_{i=1}^\infty \frac{A_{rs}^i}{u^i}, \quad (1.B.4)$$

where $\nu \ge 0$.

It follows that:

$$\frac{G_{rs}(u) - G_{rs}(\eta)}{u-\eta} = \frac{G_{rs}(u_>) - G_{rs}(u_<)}{u_> - u_<} = \sum_{i=1}^\infty \frac{A_{rs}^i}{u_>^{\nu+i} u_<^{\nu+i}} \frac{u_>^{\nu+i} - u_<^{\nu+i}}{u_> - u_<} \quad (1.B.5)$$

where $u_>$ ($u_<$) stands for the greater (smaller) between u and η.

The previous representation yields:

$$\left| \frac{G_{rs}(u) - G_{rs}(\eta)}{u-\alpha} \right|^2 = \sum_{i=1}^\infty \sum_{j=1}^\infty \frac{A_{rs}^i}{u_>^{\nu+i} u_<^{\nu+i}} \frac{(A_{rs}^j)^*}{u_>^{\nu+j} u_<^{\nu+j}} \frac{u_>^{\nu+i} - u_<^{\nu+i}}{u_> - u_<} \frac{u_>^{\nu+j} - u_<^{\nu+j}}{u_> - u_<} \quad (1.B.6)$$

Moreover, by applying the mean value theorem, we have:

$$\frac{u_>^{\nu+i} - u_<^{\nu+i}}{u_> - u_<} \le (\nu+i) u_>^{\nu+i-1} \quad (i = 1, 2 \ldots) \quad (1.B.7)$$

where the equality holds only for $\nu = 0$ and i=1 . Taking into account that

$$\begin{aligned} &(\nu+i)(\nu+j) \frac{A_{rs}^i (A_{rs}^j)^*}{u_<^{2\nu+i+j} u_>^2} + (\nu+j)(\nu+i) \frac{A_{rs}^j (A_{rs}^i)^*}{u_<^{2\nu+j+i} u_>^2} \\ &= (\nu+i)(\nu+j) \frac{A_{rs}^i (A_{rs}^j)^*}{u_<^{2\nu+i+j} u_>^2} + \left[(\nu+i)(\nu+j) \frac{A_{rs}^i (A_{rs}^j)^*}{u_<^{2\nu+i+j} u_>^2} \right]^* \end{aligned} \quad (1.B.8)$$

is real, we have:

$$\sum_{i=1}^{\infty}\sum_{j=1}^{\infty}\frac{A_{rs}^{i}}{u_{>}^{\nu+i}u_{<}^{\nu+i}}\frac{(A_{rs}^{j})^{*}}{u_{>}^{\nu+j}u_{<}^{\nu+j}}\frac{u_{>}^{\nu+i}-u_{<}^{\nu+i}}{u_{>}-u_{<}}\frac{u_{>}^{\nu+j}-u_{<}^{\nu+j}}{u_{>}-u_{<}}<$$
$$<\sum_{i=1}^{\infty}\sum_{j=1}^{\infty}(\nu+i)(\nu+j)\frac{A_{rs}^{i}(A_{rs}^{j})^{*}}{u_{<}^{2\nu+i+j}u_{>}^{2}} \tag{1.B.9}$$

It follows that:

$$\int_{M}^{\infty}\int_{M}^{\infty}\left|\frac{G_{rs}(u)-G_{rs}(\eta)}{u-\eta}\right|^{2}du\,d\alpha<\sum_{i=1}^{\infty}\sum_{j=1}^{\infty}\int_{M}^{\infty}\int_{M}^{\infty}(\nu+i)(\nu+j)\frac{A_{rs}^{i}(A_{rs}^{j})^{*}}{u_{<}^{2\nu+i+j}u_{>}^{2}}du_{<}\,du_{>}$$
$$=\sum_{i=1}^{\infty}\sum_{j=1}^{\infty}\frac{(\nu+i)(\nu+j)}{2\nu+i+j-1}\frac{A_{rs}^{i}(A_{rs}^{j})^{*}}{M^{2\nu+i+j}} \tag{1.B.10}$$

We now show that the last double series is convergent. In fact, we have:

$$\left[\left|\frac{G_{rs}(u)-G_{rs}(\eta)}{u-\eta}\right|^{2}\right]_{\substack{x=M,\\ \alpha=M}}=\left|G_{rs}'(M)\right|^{2}$$
$$=\left[\sum_{i=1}^{\infty}\sum_{j=1}^{\infty}A_{rs}^{i}(A_{rs}^{j})^{*}\frac{u^{-(\nu+i)}-\eta^{-(\nu+i)}}{u-\eta}\frac{u^{-(\nu+j)}-\eta^{-(\nu+j)}}{u-\eta}\right]_{\substack{x=M,\\ \alpha=M}} \tag{1.B.11}$$
$$=\frac{1}{M^{2}}\sum_{i=1}^{\infty}\sum_{j=1}^{\infty}(\nu+i)(\nu+j)\frac{A_{rs}^{i}(A_{rs}^{j})^{*}}{M^{2\nu+i+j}}$$

Since:

$$\frac{1}{2\nu+i+j-1}\leq 1 \tag{1.B.12}$$

it follows that:

$$\int_{M}^{\infty}\int_{M}^{\infty}\left|\frac{G_{rs}(u)-G_{rs}(\eta)}{u-\eta}\right|^{2}du\,d\alpha<\sum_{i=1}^{\infty}\sum_{j=1}^{\infty}\frac{(\nu+i)(\nu+j)}{2\nu+i+j-1}\frac{A_{rs}^{i}(A_{rs}^{j})^{*}}{M^{2\nu+i+j}}<$$
$$<\frac{1}{M^{2}}\sum_{i=1}^{\infty}\sum_{j=1}^{\infty}(\nu+i)(\nu+j)\frac{A_{rs}^{i}(A_{rs}^{j})^{*}}{M^{2\nu+i+j}}=M^{2}\left|G_{rs}'(M)\right|^{2} \tag{1.B.13}$$

This result proves that the integral $\int_{M}^{\infty}\int_{M}^{\infty}\left|\frac{[G_{rs}(u)-G_{rs}(\eta)]}{u-\eta}\right|^{2}du\,d\eta$ is bounded.

Concerning the integral $\int_{-M}^{M}\int_{M}^{\infty}\left|\frac{[G_{rs}(u)-G_{rs}(\eta)]}{u-\eta}\right|^2 du\, d\eta \;\; (-M \le u \le M, \eta \ge M)$, we have:

$$\int_{-M}^{M}\int_{M}^{\infty}\left|\frac{[G_{rs}(u)-G_{rs}(\eta)]}{u-\eta}\right|^2 du\, d\eta < \int_{M}^{\infty}\frac{1}{\eta^2}\left[\int_{-M}^{M}\left|G_{rs}(u)-G_{rs}(\eta)\right|^2 du\right]d\eta <$$
$$< \int_{M}^{\infty}\frac{1}{\eta^2}\left[\int_{-M}^{M}\left|G_{rs}(u)\right|^2 du\right]d\eta + \int_{M}^{\infty}2\frac{\left|G_{rs}(\eta)\right|}{\eta^2}\left[\int_{-M}^{M}\left|G_{rs}(u)\right| du\right]d\eta + \int_{M}^{\infty}\frac{\left|G_{rs}(\eta)\right|^2}{\eta^2}\left[\int_{-M}^{M} du\right]d\eta <$$
$$< \int_{M}^{\infty}\frac{1}{\eta^2}\left[\int_{-M}^{M}\left|G_{rs}(u)\right|^2 du\right]d\eta + \int_{M}^{\infty}2\frac{\left|G_{rs}(\eta)\right|}{\eta^2}\left[\int_{-M}^{M}\left|G_{rs}(u)\right| du\right]d\eta \tag{1.B.14}$$

Since the integrals in u are convergent and $\left|G_{rs}(\eta)\right|$ is bounded as $\eta \to \infty$, it follows that $\int_{-M}^{M}\int_{M}^{\infty}\left|\frac{[G_{rs}(x)-G_{rs}(\eta)]}{x-\eta}\right|^2 dx\, d\eta$ is also bounded.

2

A Basic Example: Scattering by a Half-plane

2.1. The fundamental problem of diffraction in wave motion

In this chapter, we will focus on a classical canonical problem that is very important in wave motion, i.e. scattering by a half-plane. We will follow, step by step, the solution of WH equations using exact closed-form methodology, and compare it with the more general and effective technique called the Fredholm factorization method, which was described in Chapter 1.

The deduction of the WH equations of this problem and the use of its solution for practical applications require some general mathematical methodologies on linear differential operators and asymptotics (Kong 1975; Felsen and Marcuvitz 1973; Daniele and Zich 2014), which will be briefly presented in this chapter. First, we will examine the electromagnetic problem and then extend the methodology to the elastic problem.

The fundamental problem of the applied electromagnetism is evaluating the electromagnetic fields $\mathcal{E}(\mathbf{r},t)$ (electrical field) and $\mathcal{H}(\mathbf{r},t)$ (magnetic field) in each point of the space $\mathbf{r}$ and in an arbitrary instant t. To take into account the presence of arbitrary media, it is convenient to add the additional fields $\mathcal{B}(\mathbf{r},t)$ (magnetic induction), $\mathcal{D}(\mathbf{r},t)$ (electric induction), $\mathcal{M}_c$ (magnetic conduction current) and $\mathcal{J}_c$ (electrical conduction current). $\mathcal{E}(\mathbf{r},t)$, $\mathcal{H}(\mathbf{r},t)$, $\mathcal{B}(\mathbf{r},t)$, $\mathcal{D}(\mathbf{r},t)$, $\mathcal{M}_c$ and $\mathcal{J}_c$ must satisfy the fundamental Maxwell equations:

This book consists of two volumes; any references to Chapters 4, 5 or 6 can be found in the second volume: *Scattering and Diffraction by Wedges 2*.

$$\nabla\times\mathcal{E}=-\frac{\partial\mathcal{B}}{\partial t}-\mathcal{M}_c-\mathcal{M}$$
$$\nabla\times\mathcal{H}=\frac{\partial\mathcal{D}}{\partial t}+\mathcal{J}_c+\mathcal{J} \tag{2.1.1}$$

where $\mathcal{J}=\mathcal{J}(\mathbf{r},t)$ (electric source current) and $\mathcal{M}=\mathcal{M}(\mathbf{r},t)$ (magnetic source current) are the known sources of the electromagnetic field.

Henceforth, we assume the presence of linear passive media and sinusoidal time sources of pulsation ω: $\mathcal{J}(\mathbf{r},t)=\mathbf{J}(\mathbf{r})e^{j\omega t}$, $\mathcal{M}(\mathbf{r},t)=\mathbf{M}(\mathbf{r})e^{j\omega t}$. It yields $\mathcal{H}(\mathbf{r},t)=\mathbf{E}(\mathbf{r})e^{j\omega t}$, $\mathcal{B}(\mathbf{r},t)=\mathbf{B}(\mathbf{r})e^{j\omega t}$, $\mathcal{D}(\mathbf{r},t)=\mathbf{D}(\mathbf{r})e^{j\omega t}$, $\mathcal{J}_c(\mathbf{r},t)=\mathbf{J}_c(\mathbf{r})e^{j\omega t}$ and $\mathcal{M}_c(\mathbf{r},t)=\mathbf{M}_c(\mathbf{r})e^{j\omega t}$.

The Maxwell equations reduce to:

$$\nabla\times\mathbf{E}=-j\omega\mathbf{B}_e-\mathbf{M}$$
$$\nabla\times\mathbf{H}=j\omega\mathbf{D}_e+\mathbf{J} \tag{2.1.2}$$

where: $\mathbf{B}_e=\mathbf{B}+\frac{1}{j\omega}\mathbf{M}_c$, $\mathbf{D}_e=\mathbf{D}+\frac{1}{j\omega}\mathbf{J}_c$.

In the following, we only consider time harmonic electromagnetic fields with a time dependence specified by the factor $e^{j\omega t}$, which is omitted.

To complete the mathematical formulation of the electromagnetic field problem, we need two additional equations. They are the constitutive relations:

$$\mathbf{D}_e=\boldsymbol{\varepsilon}\cdot\mathbf{E}+\boldsymbol{\xi}\cdot\mathbf{H}$$
$$\mathbf{B}_e=\boldsymbol{\zeta}\cdot\mathbf{E}+\boldsymbol{\mu}\cdot\mathbf{H} \tag{2.1.3}$$

The dyadics $\boldsymbol{\varepsilon},\boldsymbol{\xi},\boldsymbol{\zeta},\boldsymbol{\mu}$ are known, and specify the electromagnetic behavior of each medium present in the structure under study. In particular, for lossless media, we have the following lossless conditions (Kong 1975): $\boldsymbol{\varepsilon}=\boldsymbol{\varepsilon}^+$, $\boldsymbol{\mu}=\boldsymbol{\mu}^+$, $\boldsymbol{\zeta}=\boldsymbol{\xi}^+$, where the superscript $^+$ denotes a transpose and complex conjugate operation.

We classify materials according to their characteristic parameters $\boldsymbol{\varepsilon},\boldsymbol{\xi},\boldsymbol{\zeta},\boldsymbol{\mu}$. For example, in this book, unless otherwise specified, only homogeneous isotropic media will be considered. This means:

$$\boldsymbol{\varepsilon} = \varepsilon_r \varepsilon_o \mathbf{1}, \; \boldsymbol{\mu} = \mu_r \mu_o \mathbf{1}, \; \boldsymbol{\zeta} = 0 = \boldsymbol{\xi} \tag{2.1.4}$$

where ε_r and μ_r are, respectively, the permittivity and permeability of the material; and ε_o and μ_o are, respectively, the permittivity and permeability of the free space. In general, one or more materials in the structure under study can be dispersive in time. This means that $\varepsilon_r = \varepsilon_r(\omega)$ and/or $\mu_r = \mu_r(\omega)$.

An alternative way to model isotropic materials is based on the characterization by the phase velocity v_{ph} and the impedance Z (admittance Y=1/Z):

$$Z = \sqrt{\frac{\mu_r}{\varepsilon_r}} Z_o, \; v_{ph} = c / \sqrt{\varepsilon_r \mu_r} \tag{2.1.5}$$

where $Z_o = 377\,\Omega$ and $c = 1/\sqrt{\varepsilon_o \mu_o} = 3 \cdot 10^8 \, m/s$ are, respectively, the impedance and phase velocity of the free space. Finally, other important parameters for characterizing a material are the propagation constant k and the wavelength λ defined by:

$$k = \frac{\omega}{v_{ph}} = \omega\sqrt{\varepsilon\mu}, \; \lambda = \frac{2\pi}{k} \tag{2.1.6}$$

We observe that the imaginary part of the propagation constant k must be negative ($k = \beta - j\alpha$), which defines the selection of the sign in the square root in the presence of a complex argument in (2.1.6).

There are numerous excellent books that deal with the fundamental problem in electromagnetics. In particular, we were inspired by the treatise *Radiation and Scattering of Waves* by Felsen and Marcuvitz (1973), which intended to provide a broad methodological base that encompasses all of the specific applications. To this end, the different chapters of this book deal with structures that can be considered as waveguides with a characteristic direction. For example, we can have stratified planar, radial and angular media, and much effort has been expended on developing and applying a unified formalism for a complete analysis of these linear field problems.

As stated in the introduction, even though this systematization is debatable, in order to obtain the solution of one individual problem, for the analysis of the class of structures with similar but not identical features, it becomes essential, to eliminate redundant aspects. Therefore, transmission line formalism and, in general, equivalent circuit models, are effective tools. In fact, circuit analysis helps to order and systematize the steps necessary to obtain the equations that model the complex problem, to avoid redundancy.

In this context, we note that Felsen and Marcuvitz (1973) lack in analyzing stratified regions where discontinuities are introduced in particular sections of these regions. These authors claim that the best way to handle these problems is to use the WH technique. For example, Chapters 7 and 8 of Daniele and Zich (2014) demonstrate that the WH technique is able to solve discontinuity problems in plane stratified regions.

The main purpose of this book is to extend the WH technique to structures presenting azimuthal stratified regions, i.e. multiple wedge problems. While this aim will be accomplished in the following chapter, the scope of this chapter is to recall and apply the WH technique and the Fredholm factorization method for a basic canonical problem of wave scattering theory: the diffraction by a perfectly electrically conducting (PEC) half-plane illuminated by an Ez-polarized plane wave in electromagnetism and the diffraction by a half-infinite crack in elastic solids.

2.2. Unified theory of transverse equations in arbitrary stratified regions

Before presenting the WH technique, to solve the half-plane problem immersed in a homogeneous medium, we introduce the unified theory that considers planar discontinuities located in some sections of arbitrary stratified regions in the y-direction (Daniele and Zich 2014).

In the following, we assume Cartesian coordinates with the right-hand rule and ordering (z,x,y), i.e. $\mathbf{f}(\mathbf{r}) = \left| f_x(z,x,y), f_y(z,x,y), f_z(z,x,y) \right|$.

In each layer, the medium presents constitutive relations with the form (2.1.3). Without loss of generality, we assume the z dependence and the time dependence of the electromagnetic fields **E** and **H** specified, respectively, by the factor $e^{-j\alpha_o z}$ and by the factor $e^{j\omega t}$, which are omitted.

Using the Bresler–Marcuvitz formalism (Bresler and Marcuvitz 1956; Daniele 1971), we can deduce the transverse field equations having, as unknowns, the transverse fields $\mathbf{E}_t = \hat{z}E_z + \hat{x}E_x$ and $\mathbf{H}_t = \hat{z}H_z + \hat{x}H_x$, where $\hat{z}$, $\hat{x}$ and $\hat{y}$ are the unit vectors of the Cartesian system z,x,y.

Thus, the Maxwell equations can be rewritten in abstract form:

$$\Gamma_{\nabla} \cdot \psi = \theta \tag{2.2.1}$$

with:

$$\Gamma_{\nabla} = \begin{vmatrix} 0 & \nabla x \mathbf{1} \\ \nabla x \mathbf{1} & 0 \end{vmatrix}, \psi = \begin{vmatrix} \mathbf{E} \\ \mathbf{H} \end{vmatrix}, \theta = j\omega \begin{vmatrix} \mathbf{D} \\ -\mathbf{B} \end{vmatrix} = W \cdot \psi \tag{2.2.2}$$

where $\mathbf{1}$ is the unity dyadic in the Euclidean space and:

$$W = j\omega \begin{vmatrix} \boldsymbol{\varepsilon} & \boldsymbol{\xi} \\ -\boldsymbol{\zeta} & -\boldsymbol{\mu} \end{vmatrix} \tag{2.2.3}$$

The decomposition of

$$\nabla = \nabla_t + \hat{y}\frac{\partial}{\partial y}, \; \nabla_t = \hat{z}\frac{\partial}{\partial z} + \hat{x}\frac{\partial}{\partial x} \tag{2.2.4}$$

yields

$$\Gamma_{\nabla} = \Gamma_t + \Gamma_y \frac{\partial}{\partial y} \tag{2.2.5}$$

where:

$$\Gamma_t = \begin{vmatrix} 0 & \nabla_t x \mathbf{1} \\ \nabla_t x \mathbf{1} & 0 \end{vmatrix}, \; \Gamma_y = \begin{vmatrix} 0 & \hat{y} x \mathbf{1} \\ \hat{y} x \mathbf{1} & 0 \end{vmatrix} \tag{2.2.6}$$

The following equations hold:

$$\begin{gathered} I_t \cdot \Gamma_t = \Gamma_t \cdot I_y, I_t \cdot \Gamma_y = \Gamma_y \cdot I_t = \Gamma_y, I_y \cdot \Gamma_t = \Gamma_t \cdot I_t, \\ I_y \cdot \Gamma_y = \Gamma_y \cdot I_y = 0, \end{gathered} \tag{2.2.7}$$

where:

$$I_t = \begin{vmatrix} \mathbf{1}_t & 0 \\ 0 & \mathbf{1}_t \end{vmatrix}, I_y = \begin{vmatrix} \mathbf{1}_y & 0 \\ 0 & \mathbf{1}_y \end{vmatrix}, \tag{2.2.8}$$

$\mathbf{1}_t = \hat{z}\hat{z} + \hat{x}\hat{x}, \mathbf{1}_y = \hat{y}\hat{y}$

Taking into account these equations, the first member of (2.2.1) becomes:

$$\Gamma_\nabla \cdot \psi = (\Gamma_t + \Gamma_y \frac{\partial}{\partial y})\psi = \Gamma_t \cdot \psi_t + \Gamma_y \frac{\partial}{\partial y}\psi_t + \Gamma_t \cdot \psi_y \tag{2.2.9}$$

where $\psi_t = \begin{vmatrix} \mathbf{E}_t \\ \mathbf{H}_t \end{vmatrix}$, $\psi_y = \begin{vmatrix} E_y \hat{y} \\ H_y \hat{y} \end{vmatrix}$ with $\mathbf{E}_t = \hat{z}E_z + \hat{x}E_x$, $\mathbf{H}_t = \hat{z}H_z + \hat{x}H_x$.

Using the decomposition:

$$W = W_{tt} + W_{ty} + W_{yt} + W_{yy} \tag{2.2.10}$$

where $W_{tt} = I_t \cdot W \cdot I_t$, $W_{ty} = I_t \cdot W \cdot I_y$, $W_{yt} = I_y \cdot W \cdot I_t$, $W_{yy} = I_y \cdot W \cdot I_y$, it yields the following decomposition in transversal and longitudinal components of the Maxwell equation (2.2.1):

$$I_y \cdot \Gamma_t \cdot \psi_t = W_{yt} \cdot \psi_t + W_{yy} \cdot \psi_y \tag{2.2.11}$$

$$I_t \cdot \frac{\partial}{\partial y}\Gamma_y \cdot \psi_t + I_t \cdot \Gamma_t \cdot \psi_y = W_{tt} \cdot \psi_t + W_{ty} \cdot \psi_y \tag{2.2.12}$$

Introducing in (2.2.11) the matrix $\hat{W}_y$ defined by

$$\hat{W}_y \cdot W_{yy} = W_{yy} \cdot \hat{W}_y = I_y \tag{2.2.13}$$

it yields the equation that expresses the longitudinal field ψ_y in terms of the transversal field ψ_t:

$$\psi_y = \hat{W}_y \cdot (I_y \cdot \Gamma_t - W_{yt}) \cdot \psi_t \tag{2.2.14}$$

where:

$$\hat{W}_y = \frac{1}{j\omega(\varepsilon_y \mu_y - \xi_y \zeta_y)} \begin{vmatrix} \mu_y 1_y & \xi_y 1_y \\ -\zeta_y 1_y & -\varepsilon_y 1_y \end{vmatrix} \tag{2.2.15}$$

Taking into account that $\Gamma_y^2 = -I_t$, substituting (2.2.14) in (2.2.12) yields the transverse Maxwell equations:

$$-\frac{\partial}{\partial y}\psi_t = M(j\frac{\partial}{\partial z}, j\frac{\partial}{\partial x})\cdot\psi_t \tag{2.2.16}$$

where the matrix of order 4 $M(\frac{\partial}{\partial z}, \frac{\partial}{\partial x})$ is given by:

$$M(j\frac{\partial}{\partial z}, j\frac{\partial}{\partial x}) = -\Gamma_y\cdot\left[I_t\cdot(\Gamma_t - W_{ty})\cdot\hat{W}_y\cdot(I_y\cdot\Gamma_t - W_{yt}) - W_{tt}\right] \tag{2.2.17}$$

This operator matrix $M(j\frac{\partial}{\partial z}, j\frac{\partial}{\partial x})$ depends on ω and the parameters ε, ξ, ζ, μ of the medium. $M(\cdot,\cdot)$ is of second order in the operators $j\frac{\partial}{\partial z}$ and $j\frac{\partial}{\partial x}$.

Without loss of generality, we assume the z dependence of the electromagnetic fields **E** and **H** specified by the factor $e^{-j\alpha_o z}$, which is omitted.

Let us introduce the Fourier transforms related to ψ_t:

$$\begin{aligned}\mathbf{V}(\eta, y) &= \int_{-\infty}^{\infty} \hat{y}\times\mathbf{E}_t(\alpha_o, x, y)e^{j\eta x}dx\\ \mathbf{I}(\eta, y) &= \int_{-\infty}^{\infty} \mathbf{H}_t(\alpha_o, x, y)e^{j\eta x}dx\end{aligned} \tag{2.2.18}$$

Equation (2.2.16) yields:

$$-\frac{d}{dy}\begin{vmatrix}\mathbf{V}\\ \mathbf{I}\end{vmatrix} = \begin{vmatrix}\mathbf{T}_e(\eta) & \mathbf{Z}(\eta)\\ \mathbf{Y}(\eta) & \mathbf{T}_h(\eta)\end{vmatrix}\begin{vmatrix}\mathbf{V}\\ \mathbf{I}\end{vmatrix} \tag{2.2.19}$$

where the polynomial matrix of order 4

$$\mathbf{P}(\eta) = \begin{vmatrix}\mathbf{T}_e(\eta) & \mathbf{Z}(\eta)\\ \mathbf{Y}(\eta) & \mathbf{T}_h(\eta)\end{vmatrix} \tag{2.2.20}$$

involves the polynomial matrix of order 2 $\mathbf{T}_e(\eta), \mathbf{Z}(\eta), \mathbf{Y}(\eta), \mathbf{T}_h(\eta)$. Let us consider a homogeneous slab defined between $y = 0$ and $y = d$. The solution of (2.2.19) yields:

$$\begin{vmatrix} \mathbf{V}_1 \\ \mathbf{I}_1 \end{vmatrix} = e^{\mathbf{P}d} \begin{vmatrix} \mathbf{V}_2 \\ \mathbf{I}_2 \end{vmatrix} = \mathbf{T} \begin{vmatrix} \mathbf{V}_2 \\ \mathbf{I}_2 \end{vmatrix} \text{ or } \begin{matrix} \mathbf{V}_1 = \mathbf{A} \cdot \mathbf{V}_2 + \mathbf{B} \cdot \mathbf{I}_2 \\ \mathbf{I}_1 = \mathbf{C} \cdot \mathbf{V}_2 + \mathbf{D} \cdot \mathbf{I}_2 \end{matrix} \qquad (2.2.21)$$

where:

$$\begin{aligned} &\mathbf{V}_1 = \mathbf{V}\big|_{y=0}, \quad \mathbf{I}_1 = \mathbf{I}\big|_{y=0}, \\ &\mathbf{V}_2 = \mathbf{V}\big|_{y=d}, \quad \mathbf{I}_2 = \mathbf{I}\big|_{y=d} \end{aligned} \qquad (2.2.22)$$

and

$$\mathbf{T} = \mathbf{T}(\alpha_o, \eta) = \begin{vmatrix} \mathbf{A} & \mathbf{B} \\ \mathbf{C} & \mathbf{D} \end{vmatrix} \qquad (2.2.23)$$

is the transmission matrix of the slab. It is convenient to introduce a circuit model that represents the slab as a two-port network (Figure 2.2.1). This representation is also valid for more than one slab. In this case, the transmission matrix is the product of the transmission matrices relevant to the different slabs (Daniele and Zich 2014).

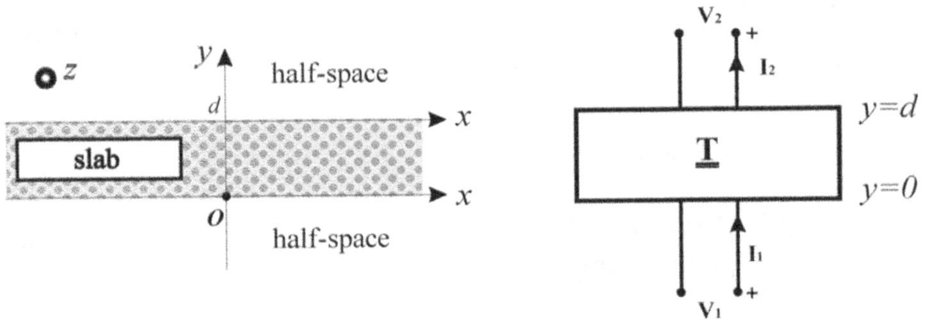

Figure 2.2.1. *Left: slab filled by an arbitrary linear medium. Right: network equivalent in the Fourier domain*

2.3. WH equation for electromagnetic diffraction by a perfectly electrically conducting (PEC) half-plane illuminated by an Ez-polarized plane wave at normal incidence

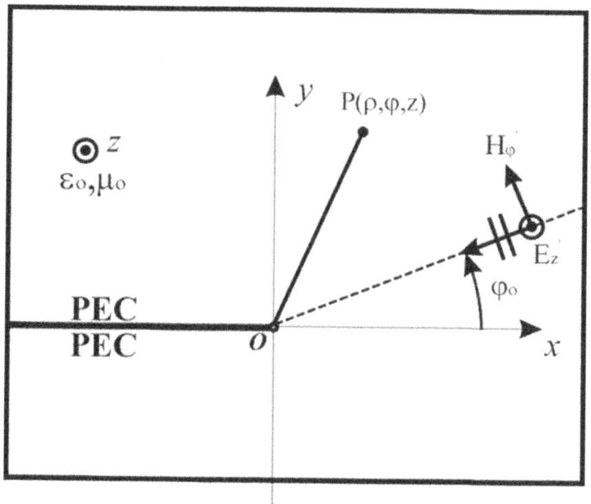

Figure 2.3.1. *The perfectly electrically conducting (PEC) half-plane problem illuminated by an Ez-polarized plane wave*

In this section, we present the WH technique to solve electromagnetic scattering by a PEC half-plane immersed in a homogeneous region (Figure 2.3.1) and illuminated by an Ez-polarized plane wave with the incident angle φ_o.

For the sake of simplicity, we assume that only the components E_z, H_x and H_y are not null, with no variations of these fields along the z-direction $(\partial / \partial z = 0)$:

$$E_z = E_z(x, y),\ H_x = H_x(x, y),\ H_y = H_y(x, y) \tag{2.3.1}$$

To deduce the WH equation of this problem, we introduce the voltage $v(\eta, y)$ and current $i(\eta, y)$ quantities in an arbitrary section y related to the spectral transformation of electromagnetic fields. These quantities are analytic functions defined by

$$\begin{aligned} v(\eta, y) &= V_-(\eta, y) + V_+(\eta, y) \\ i(\eta, y) &= I_-(\eta, y) + I_+(\eta, y) \end{aligned} \tag{2.3.2}$$

where the plus and minus Laplace transforms are the analytic continuation of the integral transformations:

$$\begin{aligned} V_-(\eta,y) &= \int_{-\infty}^{0} E_z(x,y)e^{j\eta x}dx, \quad V_+(\eta,y) = \int_{0}^{\infty} E_z(x,y)e^{j\eta x}dx \\ I_-(\eta,y) &= \int_{-\infty}^{0} H_x(x,y)e^{j\eta x}dx, \quad \mathrm{I}_+(\eta,y) = \int_{0}^{\infty} H_x(x,y)e^{j\eta x}dx \end{aligned} \tag{2.3.3}$$

When the source is a plane wave, the voltage $v(\eta,y)$ and the current $i(\eta,y)$ (as bilateral Laplace transforms, see section 1.4) satisfy the homogeneous transmission line equations (Daniele and Zich 2014):

$$\begin{aligned} -\frac{d}{dy}v(\eta,y) &= j\xi Z_c i(\eta,y) \\ -\frac{d}{dy}i(\eta,y) &= j\xi Y_c v(\eta,y) \end{aligned} \tag{2.3.4}$$

where $Z_o = \sqrt{\mu_o/\varepsilon_o}$ and $k = \omega\sqrt{\mu_o\varepsilon_o}$ are, respectively, the characteristic impedance and the propagation constant of free space.

The propagation constant ξ and the characteristic impedance Z_c (admittance $Y_c = 1/Z_c$) of the equivalent transmission line are defined by:

$$\xi = \xi(\eta) = \sqrt{k^2-\eta^2}\ , \ Z_c = Z_c(\eta) = \frac{k\,Z_o}{\xi} = \frac{k\,Z_o}{\sqrt{k^2-\eta^2}} \tag{2.3.5}$$

Taking into account the existence of only the progressive waves $e^{-j\xi y}$ as y>0 and only the regressive wave $e^{j\xi y}$ as y<0, we obtain:

$$\begin{aligned} i(\eta,0_+) &= Y_c v(\eta,0_+) \\ i(\eta,0_-) &= -Y_c v(\eta,0_-) \end{aligned} \tag{2.3.6}$$

By subtracting the above two equations and taking into account the continuity of $H_x(x,0_+) = H_x(x,0_+)$ as x>0, as well as the PEC boundary condition $E_z(x,0_+) = E_z(x,0_+) = 0$ as x<0, we have the homogeneous WH equation of the problem:

$$G(\eta)V_+(\eta) = A_-(\eta) \tag{2.3.7}$$

where:

$$G(\eta) = \frac{2\sqrt{k^2 - \eta^2}}{kZ_o} \quad (2.3.8)$$

$$A_-(\eta) = i(\eta, 0_-) - i(\eta, 0_+) = I_-(\eta, 0_-) - I_-(\eta, 0_+) \quad (2.3.9)$$

$$V_+(\eta) = v(\eta, 0_-) = v(\eta, 0_+) \quad (2.3.10)$$

$A_-(\eta)$ is a minus function that, according to the boundary condition on a PEC surface, represents the Laplace transform of the total physical current induced on both the faces of the PEC half-plane.

The advantage of using Laplace transform formalism instead of Fourier transform formalism is to avoid distributions in representing plane waves (see section 1.4 of this book and section 7.3 of (Daniele and Zich 2014)).

2.4. Non-standard contributions of WH unknowns in the PEC half-plane illuminated by an Ez-polarized plane wave at normal incidence

As illustrated in Chapter 1, the solution of the WH equations in the presence of remote sources requires the knowledge of the non-standard contributions of the WH unknowns $V_+(\eta)$ and $A_-(\eta)$ (equation (2.3.7)).

We consider the source consisting of an Ez-polarized plane wave with the incident angle φ_o:

$$E_z^i(x, y) = E_o e^{jk\rho\cos(\varphi - \varphi_o)} = E_o e^{jk(x\cos\varphi_o + y\sin\varphi_o)} \quad (2.4.1)$$

$$H_x^i(x, y) = -\frac{E_o}{Z_o} e^{jk\rho\cos(\varphi - \varphi_o)} \sin\varphi_o = -\frac{E_o}{Z_o} e^{jk(x\cos\varphi_o + y\sin\varphi_o)} \sin\varphi_o \quad (2.4.2)$$

Since the non-standard parts, when present, coincide with the non-standard part of the geometrical optical (GO) contribution $V_+^{GO}(\eta)$ and $A_-^{GO}(\eta)$, taking into account the above equations, the GO contributions in the plane y=0 are:

$$E_z^{GO}(x, 0) = E_o e^{jkx\cos\varphi_o}, \quad (x > 0) \quad (2.4.3)$$

$$H_x^{GO}(x,0_+) = -\frac{2E_o}{Z_o} e^{jkx\cos\varphi_o} \sin\varphi_o, \quad (x<0) \tag{2.4.4}$$

$$H_x^{GO}(x,0_-) = 0, \quad (x<0) \tag{2.4.5}$$

thus

$$J_z^{GO}(x) = -H_x^{GO}(x,0_+) \tag{2.4.6}$$

It yields:

$$V_+^{GO}(\eta) = j\frac{E_o}{\eta-\eta_o} \tag{2.4.7}$$

$$A_-^{GO}(\eta) = 2j\frac{E_o}{Z_o(\eta-\eta_o)}\sin\varphi_o \tag{2.4.8}$$

where $\eta_o = -k\cos\varphi_o$ is the pole of the incident wave.

To obtain the non-standard parts of $V_+^{GO}(\eta)$ and $A_-^{GO}(\eta)$, we consider the two cases $\varphi_o > \pi/2$ and $\varphi_o < \pi/2$ (with $\varphi_o = \pi/2$ being considered as a limit case that yields the pole $\eta_o = 0$ located at the real axis):

a) $\varphi_o > \pi/2 \rightarrow \mathrm{Im}[\eta_o = -k\cos\varphi_o] < 0,$

$$V_+^{n.s.}(\eta) = 0,\ A_-^{n.s}(\eta) = 2j\frac{E_o}{Z_o(\eta-\eta_o)}\sin\varphi_o$$

b) $\varphi_o < \pi/2 \rightarrow \mathrm{Im}[\eta_o = -k\cos\varphi_o] > 0$

$$V_+^{n.s.}(\eta) = j\frac{E_o}{\eta-\eta_o},\ A_-^{n.s}(\eta) = 0$$

It follows that:

$$V_+^{n.s.}(\eta) = j\frac{E_o}{\eta-\eta_o}u(\varphi_o - \frac{\pi}{2}) \tag{2.4.9}$$

$$A_-^{n.s}(\eta) = 2j\frac{E_o}{Z_o(\eta - \eta_o)}\sin\varphi_o u(-\varphi_o + \frac{\pi}{2}) \qquad (2.4.10)$$

where u(x) is the unit step Heaviside function.

2.5. Solution of the WH equation of the PEC half-plane using classical closed-form factorization

Following section 1.5, we have

$$\begin{aligned} V_+(\eta) &= [G_+(\eta)]^{-1}\left[G_+(\eta_o)V_+^{n.s.}(\eta) + G_-^{-1}(\eta_o)A_-^{n.s.}(\eta)\right] \\ A_-(\eta) &= [G(\eta)]^{-1} V_+(\eta) \end{aligned} \qquad (2.5.1)$$

where $G_+(\eta)$ is obtained using the scalar factorization:

$$G(\eta) = \frac{2\sqrt{k^2 - \eta^2}}{kZ_o} = \frac{2\xi(\eta)}{kZ_o} = G_-(\eta)G_+(\eta) \qquad (2.5.2)$$

where:

$$G_+(\eta) = \sqrt{k - \eta}\ ,\ G_-(\eta) = \frac{2}{kZ_o}\sqrt{k + \eta} \qquad (2.5.3)$$

It follows that:

$$V_+(\eta) = \frac{1}{\sqrt{k - \eta}\sqrt{k + \eta_o}}\, j\frac{kE_o \sin\varphi_o}{\eta - \eta_o} \qquad (2.5.4)$$

$$A_-(\eta) = \frac{2\sqrt{k + \eta}}{Z_o\sqrt{k + \eta_o}}\, j\frac{E_o \sin\varphi_o}{\eta - \eta_o} \qquad (2.5.5)$$

By summing equation (2.3.6), we obtain the relation between the currents $I_-(\eta, 0_-), I_-(\eta, 0_+), I_+(\eta, 0)$:

$$I_-(\eta, 0_-) + I_-(\eta, 0_+) + 2I_+(\eta, 0) = 0 \qquad (2.5.6)$$

with constant coefficients and using GO consideration for non-standard components (2.4.4)–(2.4.8), we obtain (since only the upper face of the half-plane is illuminated)

$$I_+(\eta) = I_+(\eta,0) = -\frac{I_-^{GO}(\eta,0_+)}{2} = -\frac{A_-^{GO}(\eta)}{2} = -j\frac{E_o \sin\varphi_o}{Z_o(\eta-\eta_o)} \tag{2.5.7}$$

Equation (2.5.7) represents a general result for the aperture problem of any shape (see, for example, circular apertures in (Daniele and Zich 2014)).

2.6. Solution of the WH equation of the PEC half-plane using the Fredholm factorization method

Following Chapter 1 on the FIE for the classical WH equation, the FIE for the PEC half-plane problem is

$$V_+(\eta)+\frac{[G(\eta)]^{-1}}{2\pi j}\int_{B_\theta}\frac{[G(\eta')-G(\eta)]V_+(\eta')}{\eta'-\eta}d\eta' = N_o(\eta) \qquad \eta\in B_\theta \tag{2.6.1}$$

where:

$$N_o(\eta) = [G(\eta)]^{-1} A_-^{n.s.}(\eta) + V_+^{n.s.}(\eta) \tag{2.6.2}$$

and the integration line in the η' -plane is determined to be the line defined by

$$B_\theta : \lambda_\theta(u) = e^{j\theta}u, \quad -\infty < u < \infty \tag{2.6.3}$$

The sampled version of this equation is:

$$\begin{aligned} &G(\lambda(h\,j)V_+(\lambda(h\,j))+\frac{h}{2\pi j}\cdot\sum_{i=-A/h}^{A/h} m(\lambda(h\,j),\lambda(h\,i))V_+(\lambda(h\,i))\lambda'(hi) \\ &= A_-^{n.s.}(\lambda(h\,j))+G(\lambda(h\,j))V_+^{n.s.}(\lambda(h\,j)) \end{aligned} \tag{2.6.4}$$

for $j = 0,\pm1,\pm2...,\pm(2A/h+1)$, where:

$$m(\eta,u) = If[\eta == u, G'(\eta), \frac{G(\eta)-G(u))}{\eta-u}] \tag{2.6.5}$$

In (2.6.4), the quantities h and $\pm A$ represent, respectively, the discretization step and the endpoints of the truncated integration.

Equation (2.6.4) is a linear system, whose solution provides the samples $V_+(\lambda(h\,i))$ on the integration line B_θ.

It yields the approximate representation of the plus unknown:

$$V_{+a}(\eta) = G^{-1}(\eta)\left[-\frac{h}{2\pi j}\cdot\sum_{i=-A/h}^{A/h} m(\eta,\lambda(h\,i))V_+(\lambda(h\,i))\lambda'(hi)\right] \\ +G^{-1}(\eta)A_-^{n.s.}(\eta)+V_+^{n.s.}(\eta) \tag{2.6.6}$$

For the sake of continuity, the above representation is valid until the observation point η is located in the proper sheet and does not reach any singularity line of the kernel $m(\eta,u)$ ($u \in B_\theta$ of the proper sheet). The first singularity line (for a simple scattering problem involving the classical WH equations) is defined by $\eta \in B_\theta$, with B_θ located in the improper sheet. Consequently, the above representation is valid for values of η located in the proper sheet.

To verify the validity of the approximate solution obtained using the Fredholm factorization method, we estimate the relative error:

$$e(\eta) = Abs[\frac{V_{+a}(\eta)-V_+(\eta)}{V_+(\eta)}] \tag{2.6.7}$$

We note that the relative error depends on the selection of the angle θ (2.6.3), the discretization step h and the endpoints of the truncated integration $\pm A$.

Numerical validation results are proposed in section 1.8.

2.7. Field estimation

The WH solution provides approximate spectra of $V_+(\eta)$ and $A_-(\eta)$.

The first method to estimate the field in any point of the space is based on the equivalence principle in the natural domain, together with Kirchhoff's radiation equations.

The total current $J_z(x)$ induced on the PEC half-plane is obtained using the inverse Laplace transform:

$$J_z(x) = \frac{1}{2\pi}\int_{B_-} A_-(\eta) e^{j\eta x} d\eta \tag{2.7.1}$$

where the Bromwich line B_- is a horizontal line located under all the singularities of $A_-(\eta)$.

Taking into account Green's function of two-dimensional domain, Kirchhoff's radiation equation provides the field scattered by the PEC half-plane:

$$E_z^s(x,y) = -\frac{k Z_o}{4}\int_{-\infty}^{0} H_0^{(2)}(k\left|\sqrt{(x-x')^2+y^2}\right|) J_z(x')dx' \tag{2.7.2}$$

where $H_0^{(2)}(u)$ is the Hankel function of order 0 and of second kind.

An alternative, general and more simple method to compute the field in any point of the space is based on spectral considerations, remaining in the spectral plane η.

Dealing with stratified planar structures, we introduce the transforms (2.3.2):

$$\begin{aligned} v(\eta,y) &= V_-(\eta,y) + V_+(\eta,y) \\ i(\eta,y) &= I_-(\eta,y) + I_+(\eta,y) \end{aligned} \tag{2.7.3}$$

Using the equivalence theorem on a surface at y=0 and linearity, we can decompose $v(\eta,y)$ and $i(\eta,y)$ into terms related to plus and minus equivalent currents:

$$v(\eta,y) = v^-(\eta,y) + v^+(\eta,y) \tag{2.7.4}$$

$$i(\eta,y) = i^-(\eta,y) + i^+(\eta,y) \tag{2.7.5}$$

$v^-(\eta,y)$ and $i^-(\eta,y)$ can be related to electric $(A_-(\eta))$ and magnetic $(V_-(\eta))$ surface equivalent currents distributed at x<0, y=0 according to the different formulations of the equivalence theorem, respectively, by selecting, at y=0, a perfectly magnetically conducting surface, a PEC surface or free space.

A similar interpretation holds for $v^+(\eta, y)$ and $i^+(\eta, y)$ for equivalent currents ($A_+(\eta)$ and $V_+(\eta)$) distributed at x>0, y=0.

This decomposition allows us to obtain the inverse transform of $v(\eta, y)$ and $i(\eta, y)$ (van der Pol and Bremmer 1955; Daniele and Zich 2014). The inverse transforms of $v^-(\eta, y)$ and $i^-(\eta, y)$ are

$$
\begin{aligned}
E_z^-(x,y) &= \frac{1}{2\pi}\int_{B_-} v^-(\eta,y)e^{-j\eta x}d\eta \\
H_x^-(x,y) &= \frac{1}{2\pi}\int_{B_-} i^-(\eta,y)e^{-j\eta x}d\eta
\end{aligned}
\qquad (2.7.6)
$$

where B_- is the Bromwich line associated with minus surface equivalent currents (i.e. a horizontal line located under all the singularities). Conversely, the inverse transforms of $v^+(\eta, y)$ and $i^+(\eta, y)$ are

$$
\begin{aligned}
E_z^{V_+}(x,y) &= \frac{1}{2\pi}\int_{B_+} v^+(\eta,y)e^{-j\eta x}d\eta \\
H_x^{V_+}(x,y) &= \frac{1}{2\pi}\int_{B_+} i^+(\eta,y)e^{-j\eta x}d\eta
\end{aligned}
\qquad (2.7.7)
$$

where B_+ is the Bromwich line associated with plus surface equivalent currents (i.e. a horizontal line located over all the singularities).

In particular, we apply the equivalence theorem by PEC metallization at y=0 (Figure 2.7.1), which yields

$$
v(\eta, y=0) = V_-(\eta) + V_+(\eta) \qquad (2.7.8)
$$

In this case, we reduce the evaluation of the total field in the homogeneous half-space y>0 to the sum of the primary field (the total field resulting from a flat PEC plane at y=0, i.e. the incident plus reflected field) and the radiation of the equivalent magnetic current located at the face y=0$_+$ of the PEC plane (see Figure 2.7.1).

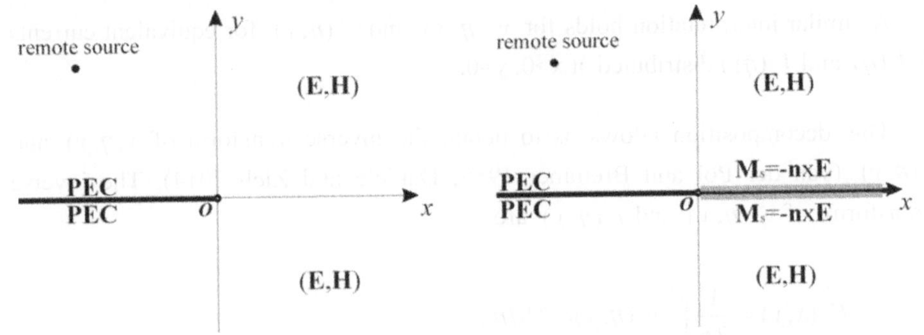

Figure 2.7.1. *Equivalence theorem using a perfectly electrically conducting surface at y=0. For a color version of the figure, see www.iste.co.uk/daniele/scattering1.zip*

Similarly, the evaluation of the total field in the homogeneous region y<0 is obtained from the radiation of the equivalent magnetic current located at the face $y=0_-$ on the PEC plane.

From (2.4.1) and (2.3.5), considering the pole of the incident wave $\eta_o = -k\cos\varphi_o$ and the propagation constant along y ($\xi(\eta) = \sqrt{k^2 - \eta^2}$), we have

$$E_z^i(x,y) = E_o e^{jk\rho\cos(\varphi-\varphi_o)} = E_o e^{jk(x\cos\varphi_o + y\sin\varphi_o)} = E_o e^{j\xi(\eta_o)y} e^{-j\eta_o x} \tag{2.7.9}$$

therefore the primary field in the half-space y>0 is

$$E_z^p(x,y) = E_z^i(x,y) - E_o e^{-j\xi(\eta_o)y} e^{-j\eta_o y} \tag{2.7.10}$$

which enforces $E_z^p(x, y=0) = 0$.

From the equivalence theorem, we have, for y>0

$$E_z(x,y) = E_z^i(x,y) - E_o e^{-j\xi(\eta_o)y} e^{-j\eta_o x} + \frac{1}{2\pi}\int_{B_+} V_+(\eta) e^{-j\xi(\eta)y} e^{-j\eta x} d\eta + \frac{1}{2\pi}\int_{B_-} V_-(\eta) e^{-j\xi(\eta)y} e^{-j\eta x} d\eta \tag{2.7.11}$$

and for y<0

$$E_z(x,y)=\frac{1}{2\pi}\int_{B_+}V_+(\eta)e^{j\xi(\eta)y}e^{-j\eta x}d\eta+\frac{1}{2\pi}\int_{B_-}V_-(\eta)e^{j\xi(\eta)y}e^{-j\eta x}d\eta \qquad (2.7.12)$$

since the primary field, in this case, is absent for the complete metallization at y=0.

Focusing on our problem (PEC half-plane), we note that only the plus magnetic surface equivalent current is non-zero, due to both the boundary condition of the problem ($V_-(\eta)=0$) and the particular application of the equivalence theorem.

Hence, the following considerations are related to the plus magnetic surface equivalent currents. However, extensions to electric/minus surface equivalent currents are straightforward.

Therefore, we need to evaluate

$$E_z^{V_+}(x,y)=\frac{1}{2\pi}\int_{B_+}V_+(\eta)e^{-j\xi(\eta)|y|}e^{-j\eta x}d\eta \qquad (2.7.13)$$

Henceforth, we will use $\bar{\eta}_o=-\cos\varphi_o$ such that $\eta_o=k\bar{\eta}_o$: $\bar{\eta}_o$ is negative when $0<\varphi_o<\pi/2$ and positive when $\pi/2<\varphi_o<\pi$. When $\bar{\eta}_o$ is negative, the pole η_o is non-standard $\mathrm{Im}[\eta_o]>0$ for the plus unknown $V_+(\eta)$.

The first step to evaluate (2.7.13) is the warping from B$_+$ to the real axis, which yields

$$E_z^{V_+}(x,y)=\frac{1}{2\pi}\int_{-\infty}^{\infty}V_+(\eta)e^{-j\xi(\eta)|y|}e^{-j\eta x}d\eta-j\,R_{V+}(\eta_o)\, e^{-j\xi(\eta_o)|y|}e^{-j\eta_o x}u[-\bar{\eta}_o] \qquad (2.7.14)$$

with $R_{V+}(\eta_o)=\mathrm{Res}[\mathrm{V}_+(\eta)]_{\eta=\eta_o}$ and where the second term related to $R_{V+}(\eta_o)$ is only present when $V_+(\eta)$ is non-standard ($u[-\bar{\eta}_o]$).

The second step is the warping of the real axis to the steepest descent path (SDP) of the integrand:

$$V_+(\eta)e^{-j\xi(\eta)|y|}e^{-j\eta x}=V_+(\eta)e^{\Omega q(\eta)} \qquad (2.7.15)$$

where:

$$q[\eta]=-j(\eta\cos\varphi+\xi(\eta)\sin\varphi) \tag{2.7.16}$$

using cylindrical coordinates ($\Omega=\rho$). The saddle point of (2.7.16) is obtained from

$$q'(\eta)=-j(\cos\varphi-\frac{\eta}{\sqrt{k^2-\eta^2}}\sin|\varphi|)=0 \tag{2.7.17}$$

which is $\eta_s=k\cos\varphi=k\bar{\eta}_s$, where $\bar{\eta}_s=\cos\varphi$ (see Figure 2.7.2 where the SDP, η_s, η_o and the vertical branch lines of $\xi(\eta)=\sqrt{k^2-\eta^2}$ are given).

At $\eta=\eta_s$

$$\begin{aligned} q(\eta_s)&=-jk \\ q''(\eta_s)&=\frac{j}{k\sin^2\varphi} \end{aligned} \tag{2.7.18}$$

The SDP is defined as the line with $\mathrm{Im}[q(\eta_s)]=\mathrm{Im}[q(\eta)]$ and $\mathrm{Re}[q(\eta_s)]\geq \mathrm{Re}[q(\eta)]$ and alternatively using $\eta=-k\cos w$ with $w=-\pi+gd(\mathrm{Im}[w])+j\,\mathrm{Im}[w]$), where $gd(x)=\mathrm{sgn}(x)\arccos(1/\cosh(x))$ is the Gudermann function.

Along the SDP, the integrand shows a fast decay with a maximum at the saddle point.

From Figure 2.7.2 we note that the pole is captured while warping the real axis to the SDP when $|\eta_o|<|\eta_s|$ or $|\bar{\eta}_o|<|\bar{\eta}_s|$. In particular, taking into account the location of the saddle point η_s and the pole η_o:

1) when $\bar{\eta}_o$ is positive, η_o is only captured clockwise if $\bar{\eta}_s>\bar{\eta}_o$ (i.e. $\mathrm{Im}[\eta_s]<\mathrm{Im}[\eta_o]$);

2) when $\bar{\eta}_o$ is negative, η_o is only captured counterclockwise if $\bar{\eta}_s<\bar{\eta}_o$ (i.e. $\mathrm{Im}[\eta_s]>\mathrm{Im}[\eta_o]$).

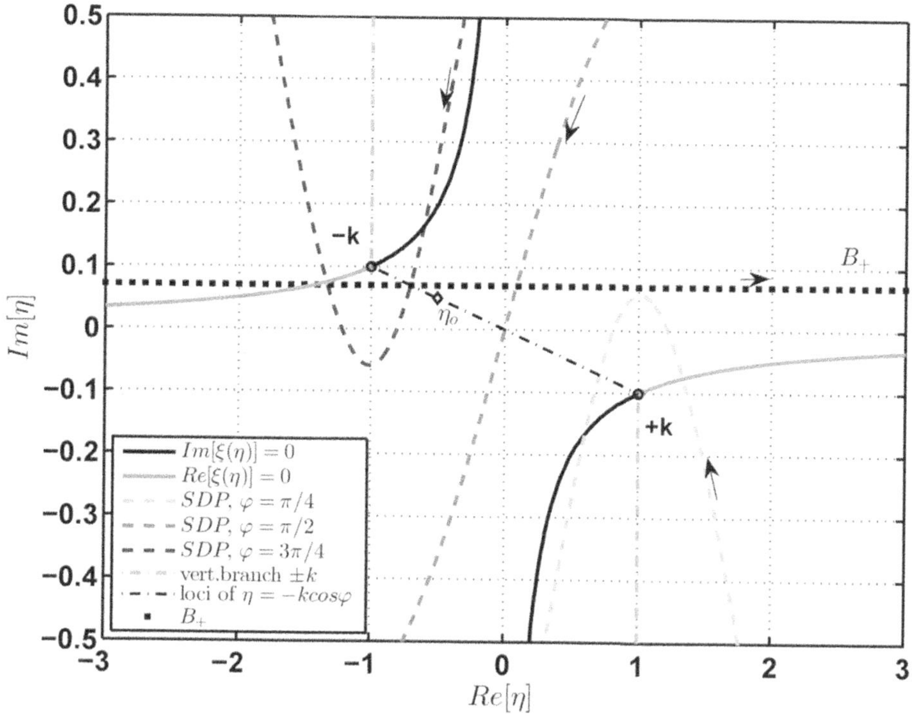

Figure 2.7.2. *Plane η: branch points $\pm k$, vertical branch lines, curves with $\mathrm{Im}[\xi(\eta)] = 0, \mathrm{Re}[\xi(\eta)] = 0$, SDP for different observation angles φ, $\eta = -k\cos\varphi$ loci of the saddle point $\eta_s = k\cos\varphi$ and of the incident plane wave pole $\eta_o = -k\cos\varphi_o$. We recall that the Bromwich contour B_+ is a horizontal line that stays above all the singularities of $V_+(\eta)$. In the case of non-standard $V_+(\eta)$, η_o is located in the upper-half of the η-plane ($\eta_o = -k\cos\varphi_o$, with $0 < \varphi_o < \pi/2$). An example of B_+ is presented in the figure for non-standard η_o while $\varphi_o = \pi/3$ (diamond symbol). An expanded version of this figure is presented in Figure 2.7.3. For a color version of the figure, see www.iste.co.uk/daniele/scattering1.zip*

Consequently:

$$I = \frac{1}{2\pi}\int_{-\infty}^{\infty} V_+(\eta)e^{-j\xi(\eta)|y|}e^{-j\eta x}d\eta + \frac{1}{2\pi}\int_{SDP} V_+(\eta)e^{-j\xi(\eta)|y|}e^{-j\eta x}d\eta =$$
$$[-j\,R_{V+}(\eta_o)u[\overline{\eta}_o]u[\overline{\eta}_s - \overline{\eta}_o] + j\,R_{V+}(\eta_o)u[-\overline{\eta}_o]u[\overline{\eta}_o - \overline{\eta}_s]]\,e^{-j\xi(\eta_o)|y|}e^{-j\eta_o x} \tag{2.7.19}$$

Taking into account that:

$$\begin{aligned}&-u[\overline{\eta}_o]u[\overline{\eta}_s-\overline{\eta}_o]+u[-\overline{\eta}_o]u[\overline{\eta}_o-\overline{\eta}_s]=\\&-u[\overline{\eta}_o]u[\overline{\eta}_s-\overline{\eta}_o]+(1-u[\overline{\eta}_o])(1-u[\overline{\eta}_s-\overline{\eta}_o])=\\&1-u[\overline{\eta}_o]-u[\overline{\eta}_s-\overline{\eta}_o]=u[-\overline{\eta}_o]-u[\overline{\eta}_s-\overline{\eta}_o]\end{aligned} \quad (2.7.20)$$

we have

$$I=j\,R_{V+}(\eta_o)(u[-\overline{\eta}_o]-u[\overline{\eta}_s-\overline{\eta}_o])e^{-j\xi(\eta_o)|y|}e^{-j\eta_o x} \quad (2.7.21)$$

Taking into account (2.7.21) in (2.7.19), it follows from (2.7.14) that

$$\begin{aligned}E_z^{V_+}(x,y)=-\frac{1}{2\pi}\int_{SDP}V_+(\eta)e^{-j\xi(\eta)|y|}e^{-j\eta x}d\eta-j\,R_{V+}(\eta_o)\\u[\overline{\eta}_s-\overline{\eta}_o]e^{-j\xi(\eta_o)|y|}e^{-j\eta_o x}\end{aligned} \quad (2.7.22)$$

In the case of the PEC half-plane problem located at x<0, y=0, we have $R_{V+}(\eta_o)=jE_o$ (incident wave contribution).

Let us now consider the integral along the SDP (Felsen and Marcuvitz 1973):

$$I_{SDP}=\frac{1}{2\pi}\int_{SDP}v(\eta)e^{-j\xi(\eta)|y|}e^{-j\eta y}d\eta=\frac{1}{2\pi}\int_{SDP}v(\eta)e^{+\Omega q(\eta)}d\eta \quad (2.7.23)$$

where, in general, we have $v(\eta)=V_-(\eta)+V_+(\eta)$ and, in the half-plane problem, $V_-(\eta)=0$.

The asymptotic evaluation of (2.7.23) is based on the change of variable:

$$s^2=q(\eta_s)-q(\eta) \quad (2.7.24)$$

The saddle point is for s=0 and the quantity $s^2=q(\eta_s)-q(\eta)$ is positive by the definition of the SDP. Figure 2.7.3 shows the mapping between the η-plane and the s-plane of the SDP, the saddle point η_s, the branches and the pole η_o.

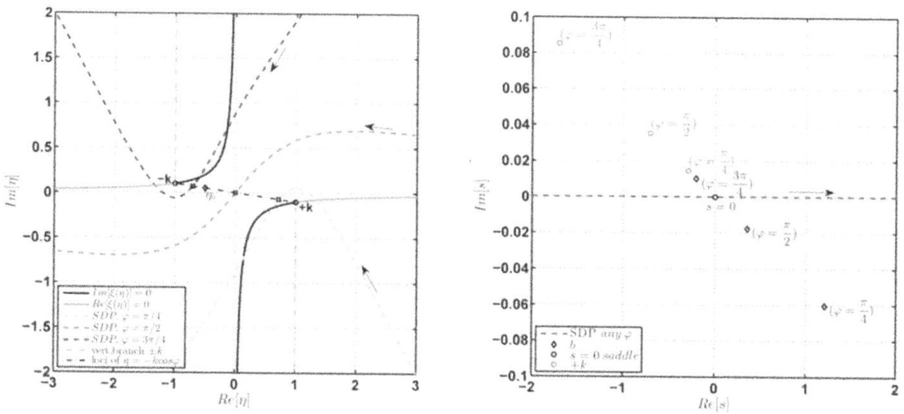

Figure 2.7.3. *Left: η-plane with the SDP for the set of $\varphi = \{\pi/4, \pi/2, 3\pi/4\}$, the saddle point $\eta_s = k\cos\varphi$ (squares), the branch point $\pm k$ (circles), the incident wave pole $\eta_o = k\cos\varphi_{os}\,(\varphi_{os} = \pi - \varphi_o)$ with $\varphi_o = \pi/3$ (diamonds). Right: s-plane (2.7.25) with the corresponding SDP curve for any φ, the pole b (2.7.31)–(2.7.36) and the branch points $+k$ for the set of $\varphi = \{\pi/4, \pi/2, 3\pi/4\}$. The location of b is in the s upper (lower) half-plane for $\varphi > \varphi_{os}$ ($\varphi < \varphi_{os}$) according to (2.7.37), which depends on both the observation angle φ and the incident plane angle φ_o. For a color version of the figure, see www.iste.co.uk/daniele/scattering1.zip*

In order to establish the sign of

$$s = \sqrt{q(\eta_s) - q(\eta)} \tag{2.7.25}$$

we observe that near the saddle point $\eta \approx \eta_s$, the Taylor expansion is

$$s = \sqrt{q(\eta_s) - q(\eta)} \approx \sqrt{-\frac{q''(\eta_s)}{2}}(\eta - \eta_s) \tag{2.7.26}$$

since $q'(\eta_s) = 0$ and thus $ds = \sqrt{-\frac{q''(\eta_s)}{2}}d\eta$. Considering (2.7.18), it yields

$$\left.\frac{d\eta}{ds}\right|_{s=0} = \pm\sqrt{-\frac{2}{q''(\eta_s)}} = \pm\sqrt{2k}e^{j\frac{\pi}{4}}\sin\varphi \tag{2.7.27}$$

taking into account that $q''(\eta_s) = j/(k\sin^2\varphi)$.

From Figure 2.7.3, we note that near the saddle point $Arg[d\eta]=\pi+\frac{\pi}{4}$, $Arg[ds]=0$. Therefore

$$\left.\frac{d\eta}{ds}\right|_{s=0}=-\sqrt{-\frac{2}{q''(\eta_s)}}=-\sqrt{2k}e^{j\frac{\pi}{4}}\sin|\varphi| \tag{2.7.28}$$

By applying the mapping (2.7.24) to (2.7.23), we obtain

$$I_{SDP}=\frac{1}{2\pi}\int_{SDP}v(\eta)e^{\Omega q(\eta)}d\eta=\frac{1}{2\pi}e^{\Omega q(\eta_s)}\int_{-\infty}^{\infty}G(s)e^{-\Omega s^2}ds=\frac{1}{2\pi}e^{-jk\rho}\int_{-\infty}^{\infty}G(s)e^{-\Omega s^2}ds \tag{2.7.29}$$

where $G(s)=v(\eta(s))\frac{d\eta}{ds}$ with $v(\eta)=V_-(\eta)+V_+(\eta)$ that shows a pole in $\eta_o=-k\cos\varphi_o=k\cos\varphi_{os}$, whose residue is

$$R=R_{V-}(\eta_o)+R_{V+}(\eta_o)=R_{V+}(\eta_o)=jE_o \tag{2.7.30}$$

The function $G(s)$ can be written as

$$G(s)=\frac{r}{s-b}+T(s) \tag{2.7.31}$$

where $T(s)$ is a smooth function on the SDP and r is the residue of $G(s)$ in $s=b$ (the corresponding point in s of η_o). Taking into account the definition of $G(s)$, we obtain that R=r:

$$\begin{aligned}&\bar{G}(s)=v(\eta(s))\\&\operatorname{Res}[\bar{G}(s)]=\operatorname{Res}[v(\eta)]ds/d\eta=Rds/d\eta\\&r=\operatorname{Res}[G(s)]=\operatorname{Res}[\bar{G}(s)]d\eta/ds=R\end{aligned} \tag{2.7.32}$$

It is important to estimate the position of the pole b in the s-plane when η_o is near η_s (i.e. b near the origin of the s-plane). Using the Taylor expansion and $q'(\eta_s)=0$, we have:

$$b^2 = q(\eta_s) - q(\eta_o) \approx -\frac{q''(\eta_s)}{2}(\eta_o - \eta_s)^2 \tag{2.7.33}$$

i.e.

$$b \approx \sqrt{-\frac{q''(\eta_s)}{2}}(\eta_o - \eta_s) \tag{2.7.34}$$

Taking into account (2.7.28), we have from (2.7.34)

$$b(\varphi) = \frac{\eta_s - \eta_o}{\sqrt{2k}\sin|\varphi|}e^{-j\frac{\pi}{4}} = \sqrt{\frac{k}{2}}\frac{\cos\varphi - \cos\varphi_{os}}{\sin|\varphi|}e^{-j\frac{\pi}{4}} \tag{2.7.35}$$

In particular, since $\mathrm{Im}[b] = -\frac{1}{2}\sqrt{k}\,\frac{\cos\varphi - \cos\varphi_{os}}{\sin|\varphi|}$

$$b(\varphi) = \frac{\eta_s - \eta_o}{\sqrt{2k}\sin|\varphi|}e^{-j\frac{\pi}{4}} = \sqrt{\frac{k}{2}}\frac{\cos\varphi - \cos\varphi_{os}}{\sin|\varphi|}e^{-j\frac{\pi}{4}} \tag{2.7.36}$$

we note that for $\varphi = \varphi_{os+} = \varphi_{os} + \varepsilon$ and $\varphi = \varphi_{os-} = \varphi_{os} - \varepsilon$ with vanishing positive ε and after the Taylor expansion, we have

$$\begin{aligned} b(\varphi_{os+}) &= -\frac{e^{-j\frac{\pi}{4}}\sqrt{k}}{\sqrt{2}}\varepsilon \\ b(\varphi_{os-}) &= \frac{e^{-j\frac{\pi}{4}}\sqrt{k}}{\sqrt{2}}\varepsilon \end{aligned} \tag{2.7.37}$$

in accordance with Figure 2.7.3.

The discontinuity of I_{SDP} (2.7.29) occurs when b crosses the saddle point s=0, i.e. $\varphi = \varphi_{os}$.

By substituting (2.7.31) into (2.7.29), we have

$$\int_{-\infty}^{\infty} G(s)e^{-\Omega s^2}ds = \int_{-\infty}^{\infty}\frac{r}{s-b}e^{-\Omega s^2}ds + \int_{-\infty}^{\infty}T(s)e^{-\Omega s^2}ds \tag{2.7.38}$$

where:

$$\int_{-\infty}^{\infty} T(s) e^{-\Omega s^2} ds = T(0)\sqrt{\frac{\pi}{\Omega}} = (G(0) + \frac{r}{b})\sqrt{\frac{\pi}{\Omega}} = (v(\eta_s)\frac{d\eta}{ds}\bigg|_{s=0} + \frac{r}{b})\sqrt{\frac{\pi}{\Omega}} \quad (2.7.39)$$

and

$$\int_{-\infty}^{\infty} \frac{r}{s-b} e^{-\Omega s^2} ds = \pm j\pi r W(\pm b\sqrt{\Omega}) \quad (2.7.40)$$

with $W(z) = e^{-z^2}(1 + \frac{2j}{\sqrt{\pi}}\int_0^z e^{-t^2} dt)$, see p. 297 of (Abramovitz and Stegun 1972).

In (2.7.40), we select + if Im[b]>0, i.e. $|\varphi| > \pi - \varphi_o$, and – if Im[b]<0, i.e. $|\varphi| < \pi - \varphi_o$. Therefore, using the definitions and computations described above, the explicit form (2.7.38) becomes

$$\int_{-\infty}^{\infty} G(s) e^{-\Omega s^2} ds = j\pi r\, sign[\varphi - \varphi_{os}] W(\sqrt{\frac{k\rho}{2}} \frac{\cos\varphi + \cos\varphi_o}{\sin|\varphi|} e^{-j\frac{\pi}{4}} sign[\varphi - \varphi_{os}]) + e^{j\frac{\pi}{4}}\sqrt{\frac{2\pi}{k\rho}}(-k\, v(\eta_s) + \frac{r}{(\cos\varphi + \cos\varphi_o)})\sin|\varphi| \quad (2.7.41)$$

Finally, for (2.7.23), we obtain the expression

$$I_{SDP} = \frac{1}{2\pi}\int_{SDP} v(\eta) e^{-j\xi(\eta)y} e^{-j\eta y} d\eta = \frac{1}{2\pi} e^{-jk\rho}\int_{-\infty}^{\infty} G(s) e^{-\Omega s^2} ds = \frac{1}{2} e^{-j(k\rho - \frac{\pi}{4})} r\, sign[\varphi - \varphi_{os}] W(\sqrt{\frac{k\rho}{2}} \frac{\cos\varphi + \cos\varphi_o}{\sin\varphi} e^{-j\frac{\pi}{4}} sign[\varphi - \varphi_{os}]) + \sqrt{\frac{1}{2\pi k\rho}} e^{-j(k\rho - \frac{\pi}{4})}\left[-k\, v(\eta_s) + \frac{r}{(\cos\varphi + \cos\varphi_o)}\right]\sin\varphi \quad (2.7.42)$$

with $r = R = jE_o$ and from (2.7.22), I_{SDP} can be interpreted as the diffractive component $E_z^{dif}(\rho,\varphi)$.

From (2.7.11) and (2.7.22), the resulting formula for the computation of the total field $E_z(x,y)$ for y>0 can be interpreted as

$$E_z(x,y) = E_z^{GO}(x,y) + E_z^d(x,y) \qquad (2.7.43)$$

where the GO component in all spaces is

$$E_z^{GO}(x,y) = E_z^i(x,y) + E_z^r(x,y) = E_z^{GO}(x,y) = E_o e^{jk\rho\cos(\varphi-\varphi_o)}[u(\varphi+\pi - \varphi_o) - u(\varphi-\pi)] - E_o e^{jk\rho\cos(\varphi+\varphi_o)}\left[u(\varphi-\pi+\varphi_o) - u(\varphi-\pi)\right] \qquad (2.7.43)$$

and the diffracted component for y>0 is $E_z^d(x,y) = -I_{SDP}$ (2.7.42). In (2.7.43), u(s) is the Heaviside step function.

The evaluation of the total field $E_z(x,y)$ for y<0 is obtained with a similar procedure taking into account that for $\varphi < 0$ the related SDP enters into the improper sheet of the η-plane. The contour integration, the estimation of $V_+(\eta)$ (2.5.4) and the singularities to be considered in this case are the ones located in the improper sheet of the η-plane.

Alternatively, as indicated for wedge problems in Chapters 3–5, we can consider the angular complex plane w (introduced in section 1.13) and the definition of rotating waves (Chapter 3) to obtain the field in any direction φ. The required spectra in this case lie in the proper sheet and the improper sheet of the η-plane mapped into the w plane. Hence, the analytic continuation of the solution obtained in the proper sheet is needed and obtained through difference equations expressed in the w plane and derived from the WH equations in η. Finally, we apply asymptotics via the SDP to obtain the total field with the GTD/UTD diffraction component.

In the next section we will demonstrate the validity of the Fredholm factorization solution for the half-plane problem by computing fields, as presented here.

2.8. Numerical validation of the Fredholm factorization method

This section has two aims:

1) to show the efficacy of the Fredholm factorization method with practical numerical results in a simple electromagnetic scattering problem (the PEC half-plane).

2) to show the convergence capabilities of the Fredholm factorization method.

First, we will show numerical results in terms of spectral unknowns, and then present the total electromagnetic field in the electromagnetic scattering problem by a plane wave.

We consider the Ez-polarized plane wave illumination (2.7.9) with incident angles $\varphi_o = 0.55\pi, 0.2\pi$, Eo=1 V/m and the propagation constant $k = (1 - j0.001)k'$, where $k' = 1$. The analysis of the problem for practical values of k' is performed by scaling the quantities. In particular, a different value of k' (e.g. $k' = p$ such that $k_{new} = p(1 - j0.001)$) changes the computed spectrum $F_+(\eta)$ to $(k / k_{new})F_+(k\eta / k_{new})$. This property can be directly ascertained by observing, in the relevant generalized WH equations, the dependence of spectral variables/unknowns and of the geometry from the propagation constant k. Alternatively, we can observe that considering any spectrum as the composition of plane waves $e^{jk\rho\cos(\varphi-\varphi_d)}$, if we apply Laplace transform (2.3.3) for different k (say k and k_{new}) to each single plane wave, we obtain the proposed relation between the spectra.

If $\varphi_o = 0.55\pi$, we have, in the formulation of the problem (2.3.7), as outlined in section 2.4, the non-standard minus function $A\_(\eta)$, while if $\varphi_o = 0.2\pi$ we have the non-standard plus function $V_+(\eta)$ (recall that $\eta_o = -k\cos\varphi_o$).

In the following, we present convergence tests for the Fredholm solution obtained in (2.6.6) through the exact analytical solution for the half-plane problem in terms of $V_+(\eta)$ described in (2.5.4). In particular, we compute the relative error in the η- and w planes ($\eta = -k\cos(w)$).

To verify the convergence, we select several sets of integration parameters A, h with $\theta = \pi/4$ in (2.6.4) and compute the relative error (2.6.7) in both the η- and w-planes.

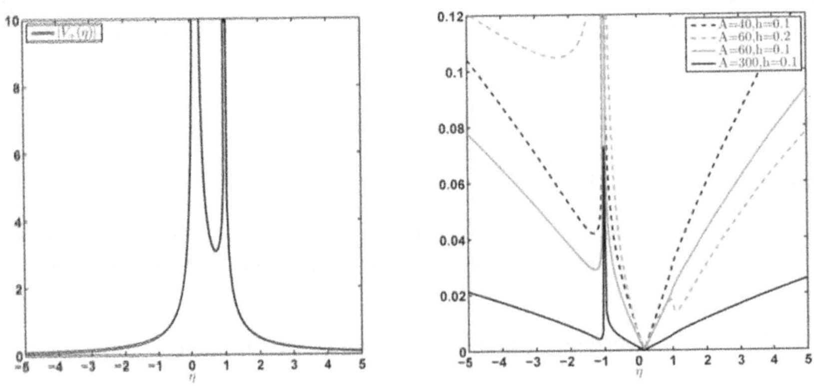

Figure 2.8.1. *The half-plane scattering problem with Eo=1 V/m, $\varphi_o = 0.55\pi$ and $k = (1 - j0.001)k'$, where $k' = 1$. Left: plots of exact solution $|V_+(\eta)|$ (2.5.4). Right: relative error (2.6.7) for different values of A and h in (2.6.4)*

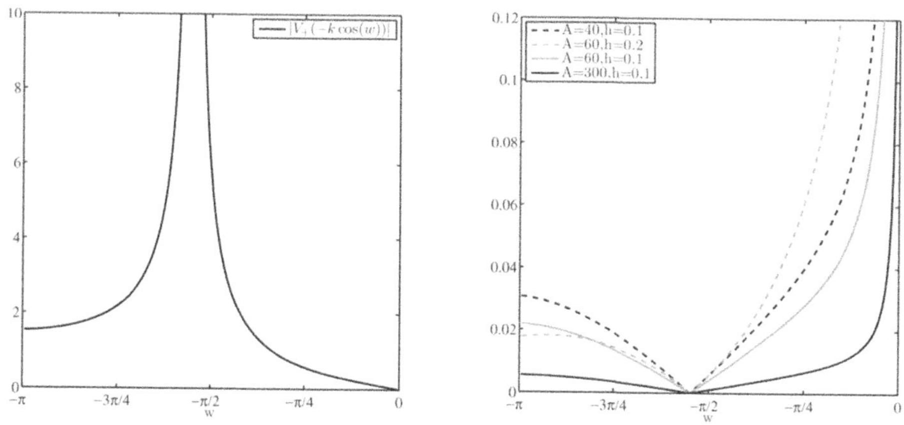

Figure 2.8.2. The half-plane scattering problem with Eo=1 V/m, $\varphi_o = 0.55\pi$ and $k = (1-j0.001)k'$, where $k' = 1$. Left: plots of exact solution $|V_+(\eta)|$ (2.5.4) in the w plane (for real w in the proper sheet, i.e. $-\pi < w < 0$). Right: relative error (2.6.7) for different values of A and h in (2.6.4) in the w plane

Figures 2.8.1 and 2.8.2 show the convergence of the Fredholm factorization solution in the η- and w planes. In particular, the numerical results demonstrate perfect agreement with the exact solution, i.e. relative errors are less than 1% for A=300, h=0.1 where the spectrum $|V_+(\eta)|$ is relevant.

We also note that, for this problem, we have selected A and h to make a compromise between computational cost and accuracy, which can be further improved by diminishing h and increasing A in order not to compromise the conditioning of the linear system. The values of A and h can be decided *a priori*, by investigating the behavior of the kernel (2.6.5) of the associated Fredholm integral equation.

For further validation, we express the solution in terms of UTD fields and total far fields.

As stated at the end of the previous section, we perform the computation using the following steps:

1) use of the w plane to represent the spectra (introduced in section 1.13);

2) analytic continuation of the solution obtained in the proper sheet through difference equations expressed in the w plane derived from the WH equations in η (section 1.16);

3) rotating waves to obtain the field spectrum in any direction φ;

4) asymptotics via the SDP to obtain the total far field with the GTD/UTD diffraction component.

These procedures for wedge problems will be presented in detail in Chapters 3–5 with appropriate references.

Here, we report the essential mathematical passages and results, since our scope is to show the validation of the Fredholm factorization method for the half-plane problem.

Following point 2 from the WH equation (2.3.7), we obtain the recursive equation (section 1.16) by eliminating the minus function based on the property that the plus function in η is an even function in w (section 1.14):

$$\hat{V}_+(w) = (\hat{G}(w))^{-1}(\hat{G}(-w-2\pi))\hat{V}_+(-w-2\pi) = -\hat{V}_+(-w-2\pi) \tag{2.8.1}$$

where $\hat{G}(w) = G(-k\cos w)$ (2.3.8) and $\hat{V}_+(w) = V_+(-k\cos w)$.

By applying rotating waves (Chapter 3), we obtain the spectrum of the electric field in the direction φ:

$$\hat{V}_{zd}(w,\varphi) = \frac{1}{2}\left[Z_o\hat{I}_+(w-\varphi) - Z_o\hat{I}(w+\varphi) + \hat{V}_d(w-\varphi) + \hat{V}_d(w+\varphi)\right] \tag{2.8.2}$$

where $\hat{V}_d(w) = \sin w\hat{V}_+(w)$ and $\hat{I}_+(w) = I_+(-k\cos w)$ with $I_+(\eta)$ given in (2.5.7) and where $\hat{V}_{zd}(w,\varphi)$ is defined as in Chapter 3, see, in particular, (3.10.3) and (3.3.2). Note that according to (2.5.7), a Fredholm factorization for this unknown has no meaning.

Finally, by applying the inverse Laplace transform with the Bromwich contour B_r and asymptotics with uniformization, we obtain the total far field:

$$E_z(\rho,\varphi) = \frac{k}{2\pi}\int_{\lambda(B_r)}\hat{V}_{zd}(w,\varphi)e^{jk_o\rho\cos w}dw = E_z^{GO}(x,y) + E_z^{UTD}(x,y) \tag{2.8.3}$$

with the GO component (2.7.43) and the uniform diffracted component:

$$E_z^{UTD}(\rho,\varphi) = E_o \frac{e^{-j(k\rho+\pi/4)}}{\sqrt{2\pi k\rho}} \left[D(\varphi,\varphi_o) - \frac{1-F\left(2k\rho\cos^2\frac{\varphi-\varphi_o}{2}\right)}{2\cos\frac{\varphi-\varphi_o}{2}} + \frac{1-F\left(2k\rho\cos^2\frac{\varphi+\varphi_o}{2}\right)}{2\cos\frac{\varphi+\varphi_o}{2}} \right] \tag{2.8.4}$$

Here, $D(\varphi,\varphi_o) = \frac{-k\hat{V}_{zd}(-\pi,\varphi)}{jE_o}$ is the GTD diffraction coefficient originated from the saddle point (w=-π) by applying the SDP method to (2.8.3). In (2.8.4), $F(z)$ is the Kouyoumjian–Pathak transition function:

$$F(z) = 2j\sqrt{z}\left[e^{jz}\int_{\sqrt{z}}^{\infty} e^{-js^2}ds - \sqrt{\pi}e^{j(z-\pi/4)}u\left(\arg\sqrt{z} - \frac{\pi}{4}\right)\right] \tag{2.8.5}$$

Figures 2.8.3 and 2.8.4 show, for different values of $\varphi_o = 0.55\pi, 0.2\pi$, the exact GO field, the UTD field component and the total far field at kρ=10, along with the relative error on the total far field at kρ=10 between the exact value and the estimation obtained using the Fredholm factorization method for different values of *A* and *h* in (2.6.4).

Both the figures show the convergence of the Fredholm factorization solution. In particular, the numerical results demonstrate perfect agreement with the exact solution, i.e. relative errors are less than 1% for *A*=300, *h*=0.1 where the total far field is relevant. Optimization on the selection of *A* and *h* should follow the criteria mentioned above concerning the spectrum. In particular, in both cases, $\varphi_o = 0.55\pi, 0.2\pi$, convergence is achieved and the properties are independent of the location of the pole $\eta_o = -k\cos\varphi_o$, which yields different non-standard components in $A_-(\eta)$, $V_+(\eta)$.

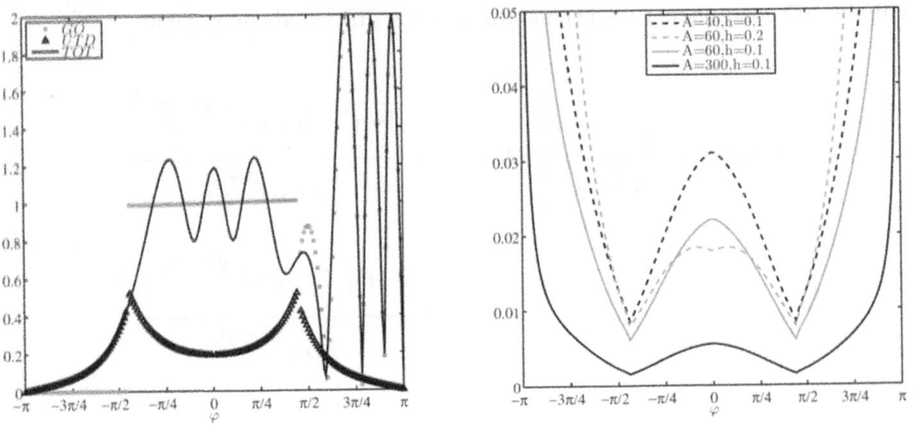

Figure 2.8.3. *Half-plane scattering problem with Eo=1 V/m, $\varphi_o = 0.55\pi$ and $k = (1 - j0.001)k'$, where $k' = 1$. Left: exact GO field, UTD field component and total far field at kρ=10. Right: relative error on the total far field at kρ=10 between the exact value and the estimation obtained using the Fredholm factorization method for different values of A and h in (2.6.4). For a color version of the figure, see www.iste.co.uk/daniele/scattering1.zip*

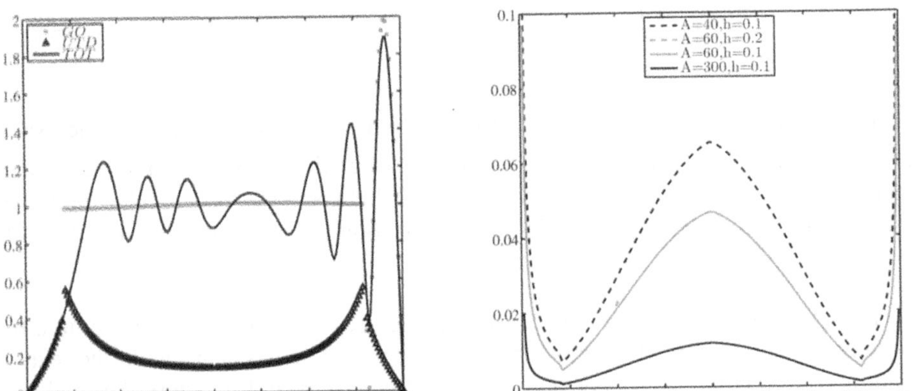

Figure 2.8.4. *Half-plane scattering problem with Eo=1 V/m, $\varphi_o = 0.2\pi$ and $k = (1 - j0.001)k'$, where $k' = 1$. Left: exact GO field, UTD field component and total far field at k=10. Right: relative error on the total far field at k=10 between the exact value and the estimation obtained using the Fredholm factorization method for different values of A and h in (2.6.4). For a color version of the figure, see www.iste.co.uk/daniele/scattering1.zip*

2.9. Generality of the wave motion in a homogeneous isotropic elastic solid

In this section, we present an extension of the WH technique used in electromagnetic scattering for solving a canonical problem in elastic solids: this problem consists of the diffraction by a half-infinite crack in an elastic solid.

Here, we only consider linear media excited by sinusoidal time sources of pulsation ω . Omitting, for simplicity, the time dependence $e^{j\omega t}$, we deal with fields that depend only on the observation point $\mathbf{r}$.

The fields, which are of interest in an elastic solid medium, are the stress tensor $\mathbf{T}=\mathbf{T}(\mathbf{r})$, the particle velocity $\mathbf{v}=\mathbf{v}(\mathbf{r})$, the strain tensor $\mathbf{S}=\mathbf{S}(\mathbf{r})$ and the momentum density $\mathbf{p}=\mathbf{p}(\mathbf{r})$. These quantities are analogous, respectively, to the electric field $\mathbf{E}$, the magnetic field $\mathbf{H}$, the dielectric induction $\mathbf{D}$ and the magnetic induction $\mathbf{B}$ (Auld 1973).

The definition of these fields can be found in any book about elasticity. We note here that the complexity of this theory requires the introduction of tensorial quantities in the 3D Euclidean space. In particular, while $\mathbf{v}=\mathbf{v}(\mathbf{r})$ and $\mathbf{p}=\mathbf{p}(\mathbf{r})$ are vectors, $\mathbf{T}=\mathbf{T}(\mathbf{r})$ and $\mathbf{S}=\mathbf{S}(\mathbf{r})$ are symmetric dyadics.

In order to maintain a strict analogy with the electromagnetic field, we only use vectorial representations of all the elastic fields.

In the following, we assume Cartesian coordinates with the right-hand rule and ordering (z,x,y) as geometry dependence, i.e. $\mathbf{f}(\mathbf{r})=|f_x(z,x,y), f_y(z,x,y), f_z(z, x,y)|$.

For example, in Cartesian coordinates, the symmetric stress tensor $\mathbf{T}=\mathbf{T}(\mathbf{r})$ that is represented by the matrix of order 3

$$\mathbf{T}=\begin{vmatrix} T_{xx} & T_{xy} & T_{xz} \\ T_{xy} & T_{yy} & T_{yz} \\ T_{xz} & T_{yz} & T_{zz} \end{vmatrix} \tag{2.9.1}$$

is redefined as a six-component column vector in Cartesian coordinates:

$$\mathbf{T} = \left|T_{xx}, T_{yy}, T_{zz}, T_{yz}, T_{xz}, T_{xy}\right|^{t} = \left|T_1, T_2, T_3, T_4, T_5, T_6\right|^{t} \tag{2.9.2}$$

where ^{t} stands for transpose.

Similarly, in Cartesian coordinates, the symmetric strain tensor $\mathbf{S} = \mathbf{S}(\mathbf{r})$ is represented by the matrix of order 3

$$\mathbf{S} = \begin{vmatrix} S_{xx} & S_{xy} & S_{xz} \\ S_{xy} & S_{yy} & S_{yz} \\ S_{xz} & S_{yz} & S_{zz} \end{vmatrix} = \begin{vmatrix} S_1 & \frac{1}{2}S_6 & \frac{1}{2}S_5 \\ \frac{1}{2}S_6 & S_2 & \frac{1}{2}S_4 \\ \frac{1}{2}S_5 & \frac{1}{2}S_4 & S_3 \end{vmatrix} \tag{2.9.3}$$

which is redefined as

$$\mathbf{S} = \left|S_1, S_2, S_3, S_4, S_5, S_6\right|^{t} \tag{2.9.4}$$

In free source regions, the field equations in an elastic medium are derived using the basic concept of mechanics, which consist of partial differential derivative equations (Auld 1973).

In particular, the analogous of the Maxwell equations assume the form:

$$\nabla_T \mathbf{T} = j\omega\mathbf{p} \tag{2.9.5}$$

$$\nabla_v \mathbf{v} = j\omega\mathbf{S} \tag{2.9.6}$$

where (2.9.5) is the equation of motion (stress–momentum relation) and (2.9.6) is the strain–velocity relation.

The differential operators given in (2.9.5) and (2.9.6) are defined as

$$\nabla_T = \begin{vmatrix} \frac{\partial}{\partial x} & 0 & 0 & 0 & \frac{\partial}{\partial z} & \frac{\partial}{\partial y} \\ 0 & \frac{\partial}{\partial y} & 0 & \frac{\partial}{\partial z} & 0 & \frac{\partial}{\partial x} \\ 0 & 0 & \frac{\partial}{\partial z} & \frac{\partial}{\partial y} & \frac{\partial}{\partial x} & 0 \end{vmatrix}, \ \nabla_v = \begin{vmatrix} \frac{\partial}{\partial x} & 0 & 0 \\ 0 & \frac{\partial}{\partial y} & 0 \\ 0 & 0 & \frac{\partial}{\partial z} \\ 0 & \frac{\partial}{\partial z} & \frac{\partial}{\partial y} \\ \frac{\partial}{\partial z} & 0 & \frac{\partial}{\partial x} \\ \frac{\partial}{\partial y} & \frac{\partial}{\partial x} & 0 \end{vmatrix} = (\nabla_T)^t \qquad (2.9.7)$$

Equations (2.9.5) and (2.9.6) can be rewritten in the same abstract form as that of the electromagnetic field presented in section 2.2:

$$\Gamma_\nabla \cdot \psi = \theta \qquad (2.9.8)$$

where:

$$\psi = \begin{vmatrix} \mathbf{T} \\ \mathbf{v} \end{vmatrix}, \ \theta = j\omega \begin{vmatrix} \mathbf{S} \\ \mathbf{p} \end{vmatrix} \qquad (2.9.9)$$

$$\Gamma_\nabla = \begin{vmatrix} 0 & \nabla_v \\ \nabla_T & 0 \end{vmatrix} \qquad (2.9.10)$$

Note that ψ is a nine-component vector (with six components in electromagnetism) and Γ_∇ is the linear differential operator of first order, according to the general theory of the abstract operator, described in section 2.2 for the electromagnetic case.

Like in electromagnetics for Maxwell equations (section 2.2), the vector equation (2.9.8) presents two vector unknowns ψ, θ and a linear differential operator of first order Γ_∇ according to the general theory of the abstract operator. Note that ψ and θ are nine-component vectors in elastics, while in electromagnetics they are six-component vectors.

In order to solve the mathematical problem, we need to enforce the constitutive relations between the field θ and ψ :

$$\theta = \mathbf{W}\ \psi \tag{2.9.11}$$

The matrix **W** depends on the physical characteristic of the medium. For the sake of simplicity, we will consider the deduction of transversal elastic equations involving only isotropic lossless material. However, we claim that the transversal elastic equations for the general case of an anisotropic medium have the same form. Therefore, in the isotropic case, we have Hooke's law in the lossless case:

$$\mathbf{T} = \mathbf{C}\mathbf{S} \tag{2.9.12}$$

and

$$\mathbf{p} = \rho\mathbf{v} \tag{2.9.13}$$

Note that (2.9.12) and (2.9.13) are, in electromagnetics, respectively equivalent to

$$\mathbf{E} = \boldsymbol{\varepsilon}^{-1}\mathbf{D} \tag{2.9.14}$$

$$\mathbf{B} = \boldsymbol{\mu}\mathbf{H} \tag{2.9.15}$$

Their abstract form defines the following matrix W:

$$\mathbf{W} = \begin{vmatrix} \mathbf{0} & \boldsymbol{\rho} \\ \mathbf{C}^{-1} & \mathbf{0} \end{vmatrix} \tag{2.9.16}$$

where:

$$\mathbf{C} = \begin{vmatrix} \lambda+2\mu & \lambda & \lambda & 0 & 0 & 0 \\ \lambda & \lambda+2\mu & \lambda & 0 & 0 & 0 \\ \lambda & \lambda & \lambda+2\mu & 0 & 0 & 0 \\ 0 & 0 & 0 & \mu & 0 & 0 \\ 0 & 0 & 0 & 0 & \mu & 0 \\ 0 & 0 & 0 & 0 & 0 & \mu \end{vmatrix}, \quad \boldsymbol{\rho} = \begin{vmatrix} \rho & 0 & 0 \\ 0 & \rho & 0 \\ 0 & 0 & \rho \end{vmatrix} \tag{2.9.17}$$

The parameters that characterize the isotropic material are the density ρ of the medium and Lame's elastic constants λ and μ.

Lame's elastic constants are dimensionally analogous to the reciprocal of the electric permittivity $\boldsymbol{\varepsilon}^{-1}$, and the density ρ is analogous to the magnetic permeability μ_m. Consequently, the electromagnetic wavenumber $\omega\sqrt{\mu_m \varepsilon}$ has elastic analogs of same dimension $\omega\sqrt{\rho/\lambda}$, see (2.10.13)–(2.10.14). Moreover, the electromagnetic impedance (electric field/magnetic field ratio) $\sqrt{\mu_m/\varepsilon}$ has elastic analogs of same dimension $\sqrt{\rho\lambda}$, see the stress/velocity ratio using the components of (2.10.17). We note that the dyadic isotropic elastic problem is analogous to the anisotropic electromagnetic problem.

The presence of losses in the material is due to viscoelastic damping. It makes the medium dispersive by modifying Hooke's law into the form (Auld 1973):

$$\mathbf{T} = (\mathbf{C} + j\omega\boldsymbol{\eta})\mathbf{S} \tag{2.9.18}$$

where the matrix $\boldsymbol{\eta}$ depends on the material and temperature and vanishes in the absence of viscoelastic damping.

By introducing the constitutive equations, the study of the elastic motion in a material requires the solution of the elasticity equations that in the absence of external sources assume the form:

$$(\Gamma_{\text{V}} - j\omega\mathbf{W})\psi = 0 \tag{2.9.19}$$

If the considered material is fluid, the stress tensor becomes a scalar (pressure with an inverted sign) and the Lame parameter μ is vanishing. Consequently, the vectors θ and ψ and the matrices Γ_{V} and $\mathbf{W}$ are of dimension 4 instead of 9. In particular, the fluid is characterized by only two parameters (λ and ρ) instead of three (λ, μ and ρ) that are introduced in a solid.

As in the electromagnetic field, in the presence of stratified media (assuming stratification along y), it is convenient to introduce the transversal equation of the elasticity equations. We consider the general theory of the transversalization of arbitrary field equations arranged in systems of partial differential equations of the first order (Bresler and Marcuvitz 1956; Daniele 1971).

Therefore, it is important to distinguish the continuous component from the discontinuous one in the elastic field $\psi = \begin{vmatrix} \mathbf{T} \\ \mathbf{v} \end{vmatrix}$ (2.9.9). While the stratification is in the y-direction, using the divergence theorem (Daniele 1971), we demonstrate the continuity of the velocity vector $\mathbf{v}$ and the vector $\hat{y}\cdot\mathbf{T}$, i.e. the scalar quantities represented by all the components of $\mathbf{v} = |v_x, v_y, v_z|^t = |v_1, v_2, v_3|^t$ and the even components of the stress vector T (2.9.2), i.e. $v_1, v_2, v_3, T_2, T_4, T_6$.

The wave vector ψ can be rewritten as

$$\psi = \begin{vmatrix} \psi_c \\ \psi_d \end{vmatrix} \tag{2.9.20}$$

where the vector of order 6

$$\psi_c = |T_2, T_4, T_6, v_1, v_2, v_3|^t \tag{2.9.21}$$

and the vector of order 3

$$\psi_d = |T_1, T_3, T_5|^t \tag{2.9.22}$$

are, respectively, continuous and discontinuous in the stratification along the y-direction.

The procedure to get the transversal equations for the general case requires some effort. Alternatively, while considering isotropic elastic media, the direct deduction is very simple.

For example, we use the explicit form of (2.9.8) and (2.9.11). After redefining tensor quantities as vectors, we obtain a system of nine equations $\Gamma_\nabla \psi = j\omega W \psi$ (2.9.19), i.e.

$$\begin{cases} D_xT_1 + D_zT_5 + D_yT_6 = j\rho\omega v_1 \\ D_yT_2 + D_zT_4 + D_xT_6 = j\rho\omega v_2 \\ D_zT_3 + D_yT_4 + D_xT_5 = j\rho\omega v_3 \\ D_xv_1 = j(-\dfrac{2T_2\lambda\mu}{12\lambda\mu^2+8\mu^3} - \dfrac{2T_3\lambda\mu}{12\lambda\mu^2+8\mu^3} + \dfrac{T_1(4\lambda\mu+4\mu^2)}{12\lambda\mu^2+8\mu^3})\omega \\ D_yv_2 = j(-\dfrac{2T_1\lambda\mu}{12\lambda\mu^2+8\mu^3} - \dfrac{2T_3\lambda\mu}{12\lambda\mu^2+8\mu^3} + \dfrac{T_2(4\lambda\mu+4\mu^2)}{12\lambda\mu^2+8\mu^3})\omega \\ D_zv_3 = j(-\dfrac{2T_1\lambda\mu}{12\lambda\mu^2+8\mu^3} - \dfrac{2T_2\lambda\mu}{12\lambda\mu^2+8\mu^3} + \dfrac{T_3(4\lambda\mu+4\mu^2)}{12\lambda\mu^2+8\mu^3})\omega \\ D_zv_2 + D_yv_3 = \dfrac{jT_4\omega}{\mu} \\ D_zv_1 + D_xv_3 = \dfrac{jT_5\omega}{\mu} \\ D_yv_1 + D_xv_2 = \dfrac{jT_6\omega}{\mu} \end{cases} \tag{2.9.23}$$

where $D_x = \dfrac{\partial}{\partial x}, D_y = \dfrac{\partial}{\partial y}, D_z = \dfrac{\partial}{\partial z}$.

From the fourth, sixth and eighth lines of (2.9.23), we obtain the odd components of T (i.e. ψ_d) in terms of the even components of T and the components of v (i.e. ψ_c):

$$\begin{cases} T_1 = -\dfrac{j(4D_xv_1\lambda\mu + 2D_zv_3\lambda\mu + 4D_xv_1\mu^2 + jT_2\lambda\omega)}{(\lambda+2\mu)\omega} \\ T_3 = -\dfrac{j(2D_xv_1\lambda\mu + 4D_zv_3\lambda\mu + 4D_zv_3\mu^2 + jT_2\lambda\omega)}{(\lambda+2\mu)\omega} \\ T_5 = -\dfrac{j(D_zv_1 + D_xv_3)\mu}{\omega} \end{cases} \tag{2.9.24}$$

We observe that (2.9.24) does not depend on the derivative Dy in the continuous field ψ_c.

By substituting (2.9.24) into the 6 non-used equations of (2.9.23) (i.e. equations at lines 1, 2, 3, 5, 7 and 9), we obtain the required transverse equations:

$$-\frac{\partial}{\partial y}\psi_c = M(j\frac{\partial}{\partial z}, j\frac{\partial}{\partial x})\cdot\psi_c \tag{2.9.25}$$

where the matrix $M(j\frac{\partial}{\partial z}, j\frac{\partial}{\partial x})$ is of dimension 6 (which is not explicitly reported here) with $\psi_c = |T_2, T_4, T_6, v_1, v_2, v_3|^t$. We observe that in electromagnetic problems, ψ_c consists of 4 tangential components of the electric and magnetic fields.

We also claim that in the case of the general anisotropic elastic problem, the transverse equations assume the same form as (2.9.25) with more complicated parameters.

Without loss of generality, in geometries that are invariant in the z-direction, we assume the z dependence of the elastic field **T** and **v** specified by the factor $e^{-j\alpha_o z}$, which is omitted. To enforce the analogy with the wave motion of electromagnetic fields in stratified regions, we define the spectral transformation along x:

$$\mathbf{V}(\eta, y) = \int_{-\infty}^{\infty}\begin{vmatrix} T_2(\alpha_o, x, y) \\ T_4(\alpha_o, x, y) \\ T_6(\alpha_o, x, y)\end{vmatrix} e^{j\eta x}dx, \quad \mathbf{I}(\eta, y) = \int_{-\infty}^{\infty}\begin{vmatrix} v_1(\alpha_o, x, y) \\ v_2(\alpha_o, y, x) \\ v_3(\alpha_o, x, y)\end{vmatrix} e^{j\eta x}dx \tag{2.9.26}$$

In electromagnetics, $\mathbf{V}(\eta, y)$ and $\mathbf{I}(\eta, y)$ are, respectively, the Fourier transforms of the component of the electric and magnetic fields located in the z-x plane transverse to the stratification along y.

With these notations, it is possible for the elastic wave motion to apply the general theory in stratified planar regions, as discussed in (Daniele and Zich 2014). In fact, we also take into account that in the presence of an arbitrary homogeneous elastic medium, equation (2.9.25) yields:

$$-\frac{d}{dy}\begin{vmatrix}\mathbf{V} \\ \mathbf{I}\end{vmatrix} = \mathbf{P}(\alpha_o, \eta)\begin{vmatrix}\mathbf{V} \\ \mathbf{I}\end{vmatrix} \tag{2.9.27}$$

where the polynomial matrix of order 2

$$\mathbf{P}(\alpha_o,\eta) = \begin{vmatrix} \mathbf{T}_e(\eta) & \mathbf{Z}(\eta) \\ \mathbf{Y}(\eta) & \mathbf{T}_h(\eta) \end{vmatrix} = \mathbf{M}(\alpha_o,\eta) \tag{2.9.28}$$

has matrix components $\mathbf{T}_e(\eta), \mathbf{Z}(\eta), \mathbf{Y}(\eta), \mathbf{T}_h(\eta)$ of order 3 and $\begin{vmatrix} \mathbf{V} \\ \mathbf{I} \end{vmatrix}$ is the spectral transformation along x of ψ_c (in the following we will use the same symbol ψ_c for the quantity in both the geometrical and spectral domains).

Let us consider a homogeneous slab defined between $y = 0$ and $y = d$. The solution of (2.9.27) yields:

$$\begin{vmatrix} \mathbf{V}_0 \\ \mathbf{I}_0 \end{vmatrix} = e^{\mathbf{P}d} \begin{vmatrix} \mathbf{V}_d \\ \mathbf{I}_d \end{vmatrix} \tag{2.9.29}$$

which can be rewritten, using the transmission matrix notation:

$$\begin{vmatrix} \mathbf{V}_0 \\ \mathbf{I}_0 \end{vmatrix} = \mathcal{T} \begin{vmatrix} \mathbf{V}_d \\ \mathbf{I}_d \end{vmatrix} \tag{2.9.30}$$

where:

$$\begin{aligned} \mathbf{V}_0 &= \mathbf{V}\big|_{y=0}, \quad \mathbf{I}_0 = \mathbf{I}\big|_{y=0} \\ \mathbf{V}_d &= \mathbf{V}\big|_{y=d}, \quad \mathbf{I}_d = \mathbf{I}\big|_{y=d} \end{aligned} \tag{2.9.31}$$

$$\mathcal{T} = \mathcal{T}(\alpha_o,\eta) = \begin{vmatrix} \mathbf{A} & \mathbf{B} \\ \mathbf{C} & \mathbf{D} \end{vmatrix} = e^{\mathbf{P}d} \tag{2.9.32}$$

Here, $\mathcal{T}$ is the transmission matrix of the slab. It is convenient to introduce a two-port network model representing the slab (Figure 2.9.1). This representation is valid for consecutive slabs. In this case, the transmission matrix of the multi-layer slab is the product of the transmission matrices relevant to the different slabs (Daniele and Zich 2014) (Figure 2.9.2).

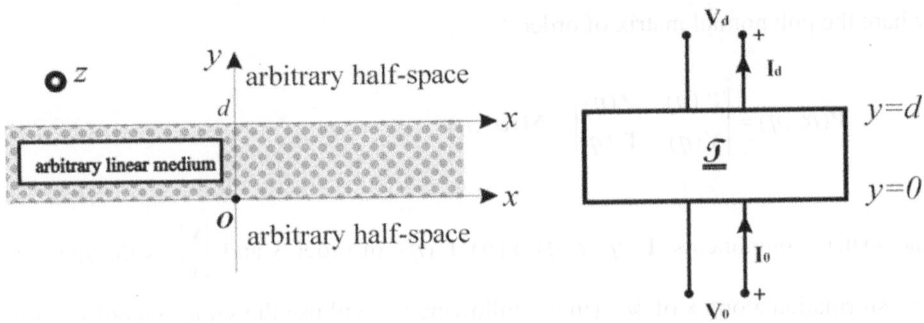

Figure 2.9.1. *Left: slab filled by an arbitrary linear medium in elasticity. Right: network equivalent in the Fourier domain*

If the slab is indefinite in the positive (or negative) y-direction, the vanishing of $\mathbf{V_d}$ and $\mathbf{I_d}$ as $d \to \infty$ is compensated by increasing the matrix $\mathcal{T}$ so that it is necessary to introduce a limit process in order to model a homogeneous semi-infinite medium. As this process can be complex, it is better to introduce the eigenvalues and eigenvectors of the matrix $\mathbf{P}$ according to the following considerations.

By indicating the six eigenvalues and eigenvectors of $\mathbf{P}$ with $\gamma_i = \gamma_i(\eta)$, $i = 1,2,3,4,5,6$ and $\psi_c^{(i)} = \begin{vmatrix} \mathbf{V}^{(i)} \\ \mathbf{I}^{(i)} \end{vmatrix}$, $i = 1,2,3,4,5,6$, the general solution of (2.9.27) assumes the form:

$$\begin{vmatrix} \mathbf{V} \\ \mathbf{I} \end{vmatrix} = C_1 e^{-\gamma_1 y} \psi_c^{(1)} + C_2 e^{-\gamma_2 y} \psi_c^{(2)} + C_3 e^{-\gamma_3 y} \psi_c^{(3)} + C_4 e^{-\gamma_4 y} \psi_c^{(4)} + C_5 e^{-\gamma_5 y} \psi_c^{(5)} + C_6 e^{-\gamma_6 y} \psi_c^{(6)} \quad (2.9.33)$$

i.e.

$$\begin{cases} \mathbf{V} = C_1 e^{-\gamma_1 y} \mathbf{V}^{(1)} + C_2 e^{-\gamma_2 y} \mathbf{V}^{(2)} + C_3 e^{-\gamma_3 y} \mathbf{V}^{(3)} + C_4 e^{-\gamma_4 y} \mathbf{V}^{(4)} \\ \qquad + C_5 e^{-\gamma_5 y} \mathbf{V}^{(5)} + C_6 e^{-\gamma_6 y} \mathbf{V}^{(6)} \\ \mathbf{I} = C_1 e^{-\gamma_1 y} \mathbf{I}^{(1)} + C_2 e^{-\gamma_2 y} \mathbf{I}^{(2)} + C_3 e^{-\gamma_3 y} \mathbf{I}^{(3)} + C_4 e^{-\gamma_4 y} \mathbf{I}^{(4)} + C_5 e^{-\gamma_5 y} \mathbf{I}^{(5)} + C_6 e^{-\gamma_6 y} \mathbf{I}^{(6)} \end{cases} \quad (2.9.34)$$

where C_i, $i = 1,2,3,4,5,6$ are six arbitrary scalars that do not depend on y.

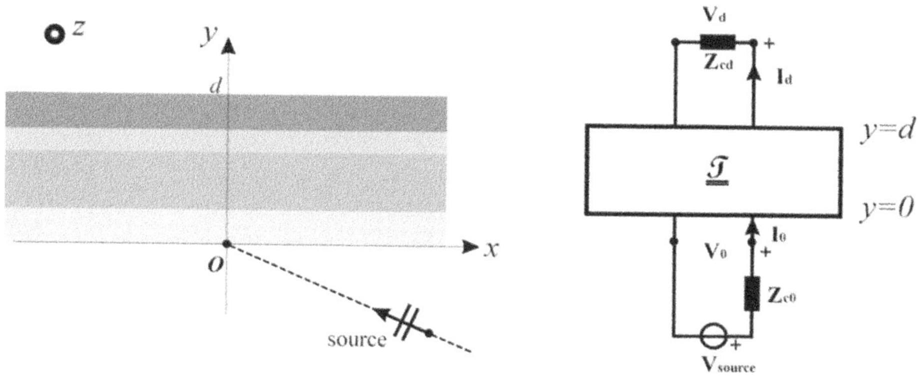

Figure 2.9.2. *Left: multilayer slab in elasticity terminated by semi-infinite regions and illuminated by a plane wave from the bottom. Right: network modeling of the multilayer slab where the transmission matrix of the multi-layer slab is the product of the transmission matrices relevant to the different slabs and the homogeneous half-spaces are modeled by Thevenin models as described in section 2.10 for plane waves*

In the presence of a passive medium, we conjecture that three eigenvalues $\gamma_i = \gamma_i(\eta)$ (e.g. $\gamma_1, \gamma_2, \gamma_3$) present a non-negative real part and the other two (e.g. $\gamma_4, \gamma_5, \gamma_6$) present a non-positive real part. It follows that in an indefinite upper medium, the scalars C_4, C_5, C_6 must be zero and the solution must have the form

$$\begin{cases} \mathbf{V} = C_1 e^{-\gamma_1 y}\mathbf{V}^{(1)} + C_2 e^{-\gamma_2 y}\mathbf{V}^{(2)} + C_3 e^{-\gamma_3 y}\mathbf{V}^{(3)} \\ \mathbf{I} = C_1 e^{-\gamma_1 y}\mathbf{I}^{(1)} + C_2 e^{-\gamma_2 y}\mathbf{I}^{(2)} + C_3 e^{-\gamma_3 y}\mathbf{I}^{(3)} \end{cases} \quad y > 0 \qquad (2.9.35)$$

Conversely, in the presence of an indefinite lower medium, the scalars C_1, C_2, C_3 must be zero and the solution must have the form

$$\begin{cases} \mathbf{V} = C_4 e^{-\gamma_4 y}\mathbf{V}^{(4)} + C_5 e^{-\gamma_5 y}\mathbf{V}^{(5)} + C_6 e^{-\gamma_6 y}\mathbf{V}^{(6)} \\ \mathbf{I} = C_4 e^{-\gamma_4 y}\mathbf{I}^{(4)} + C_5 e^{-\gamma_5 y}\mathbf{I}^{(5)} + C_6 e^{-\gamma_6 y}\mathbf{I}^{(6)} \end{cases} \quad y < 0 \qquad (2.9.36)$$

We can electromagnetically interpret (2.9.33) and (2.9.34), respectively, as a generalization of progressive and regressive propagation in a linear medium. However, while in bi-anisotropic electromagnetic media only two waves (ordinary and extraordinary waves) are present in each direction, in elastic media there are three. In particular, in isotropic elastic media, the three elastic waves in (2.9.33) and

(2.9.34) represent the longitudinal wave and the two shear waves (the vertical and horizontal waves).

The components of the eigenvectors can be written in the form

$$\mathbf{V}^{(i)} = \begin{vmatrix} V_1^{(i)} \\ V_2^{(i)} \\ V_3^{(i)} \end{vmatrix}, \mathbf{I}_i = \begin{vmatrix} I_1^{(i)} \\ I_2^{(i)} \\ I_3^{(i)} \end{vmatrix} \quad i = 1,2,3,4,5,6 \tag{2.9.37}$$

It follows for y>0:

$$\mathbf{V} = \begin{vmatrix} V_1 \\ V_2 \\ V_3 \end{vmatrix} = \begin{vmatrix} V_1^{(1)} & V_1^{(2)} & V_1^{(3)} \\ V_2^{(1)} & V_2^{(2)} & V_2^{(3)} \\ V_3^{(1)} & V_3^{(2)} & V_3^{(3)} \end{vmatrix} \begin{vmatrix} C_1 e^{-\gamma_1 y} \\ C_2 e^{-\gamma_2 y} \\ C_3 e^{-\gamma_3 y} \end{vmatrix} \tag{2.9.38}$$

$$\mathbf{I} = \begin{vmatrix} I_1 \\ I_2 \\ I_3 \end{vmatrix} = \begin{vmatrix} I_1^{(1)} & I_1^{(2)} & I_1^{(3)} \\ I_2^{(1)} & I_2^{(2)} & I_2^{(3)} \\ I_3^{(1)} & I_3^{(2)} & I_3^{(3)} \end{vmatrix} \begin{vmatrix} C_1 e^{-\gamma_1 y} \\ C_2 e^{-\gamma_2 y} \\ C_3 e^{-\gamma_3 y} \end{vmatrix} \tag{2.9.39}$$

By eliminating the vector $\begin{vmatrix} C_1 e^{-\gamma_1 y} \\ C_2 e^{-\gamma_2 y} \\ C_3 e^{-\gamma_3 y} \end{vmatrix}$, (2.9.38) and (2.9.39) yield:

$$\mathbf{V} = \vec{\mathbf{Z}}_c \mathbf{I}, \ y > 0 \tag{2.9.40}$$

with

$$\vec{\mathbf{Z}}_c = \begin{vmatrix} V_1^{(1)} & V_1^{(2)} & V_1^{(3)} \\ V_2^{(1)} & V_2^{(2)} & V_2^{(3)} \\ V_3^{(1)} & V_3^{(2)} & V_3^{(3)} \end{vmatrix} \cdot \begin{vmatrix} I_1^{(1)} & I_1^{(2)} & I_1^{(3)} \\ I_2^{(1)} & I_2^{(2)} & I_2^{(3)} \\ I_3^{(1)} & I_3^{(2)} & I_3^{(3)} \end{vmatrix}^{-1} \tag{2.9.41}$$

Equations (2.9.40) and (2.9.41) extend the fundamental concept of the characteristic impedance $\vec{\mathbf{Z}}_c$ to an indefinite arbitrary linear medium for y>0.

Similarly, for an indefinite lower medium y<0, we have:

$$\mathbf{V} = -\overleftrightarrow{\mathbf{Z}}_c \, \mathbf{I}\,, \; y<0 \tag{2.9.42}$$

where the characteristic impedance $\overleftrightarrow{\mathbf{Z}}_c$ is defined by:

$$\overleftrightarrow{\mathbf{Z}}_c = -\begin{vmatrix} V_1^{(4)} & V_1^{(5)} & V_1^{(6)} \\ V_2^{(4)} & V_2^{(5)} & V_2^{(6)} \\ V_3^{(4)} & V_3^{(5)} & V_3^{(6)} \end{vmatrix} \cdot \begin{vmatrix} I_1^{(4)} & I_1^{(5)} & I_1^{(6)} \\ I_2^{(4)} & I_2^{(5)} & I_2^{(6)} \\ I_3^{(4)} & I_3^{(5)} & I_3^{(6)} \end{vmatrix}^{-1} \tag{2.9.43}$$

2.10. Plane waves in a homogeneous elastic solid and simplifications

Plane waves in an indefinite homogeneous elastic solid medium are defined by

$$\psi = \psi_o e^{-jk_x x - jk_y y - jk_z z} = \psi_o e^{-\mathbf{k}\cdot\mathbf{r}} \tag{2.10.1}$$

where ψ_o is a constant vector of order 9 according to (2.9.20) and where the wave vector is $\mathbf{k} = k\hat{\mathbf{k}} = k_x\hat{x} + k_y\hat{y} + k_z\hat{z}$. Taking into account that

$$\frac{\partial}{\partial x} = -jk_x \quad \frac{\partial}{\partial y} = -jk_y \quad \frac{\partial}{\partial z} = -jk_z \tag{2.10.2}$$

equation (2.9.25) becomes:

$$jk_y \psi_c = \mathbf{M}(k_z, k_x) \cdot \psi_c \tag{2.10.3}$$

as defined in (2.9.27)–(2.9.28). Equation (2.10.3) is of order 6, which determines the plane wave spectra in the medium in terms of k_x, k_y, k_z.

Here, the computation is performed for a particular plane wave instead of performing Fourier transformation along x, as in (2.9.26). We assume an $e^{-j\alpha_o z} e^{-j\eta_o x}$ dependence that defines a specific direction of incidence: $\eta_o = -\tau_o \cos(\varphi_o)$, $\alpha_o = k\cos\beta$, $\tau_o = k\sin\beta = \sqrt{k^2 - \alpha_o^2}$, β and φ_o are respectively the zenithal (skewness) and the azimuthal angles in spherical coordinates. We recall that when

Fourier transformation is applied along x, we have $e^{-j\eta_o x} \rightarrow 2\pi\delta(\eta-\eta_o)$. However, in the following, we specialize the procedure for only a plane wave of the form

$$\psi = \psi_o e^{-j\eta_o x - jk_y y - j\alpha_o z} \tag{2.10.4}$$

in order to avoid distribution formalism.

From (2.10.3) and (2.10.4), it is evident that the dispersion equation is a scalar equation of order 6 that is obtained from the characteristic polynomial

$$\det[jk_y \mathrm{I} - \mathbf{M}(\alpha_o, \eta_o)] = 0 \tag{2.10.5}$$

where $k_z = \alpha_o$, $k_x = \eta_o$ and I is the identity matrix of order 6. The six solutions of (2.10.5) are

$$jk_y^i = \gamma_i(\alpha_o, \eta_o) \qquad \text{i=1,2,..,6} \tag{2.10.6}$$

where γ_i are the eigenvalues of the matrix $\mathbf{M}(\alpha_o, \eta_o) \equiv \mathbf{P}(\alpha_o, \eta_o)$ with related eigenvectors

$$\psi_c^i = \psi_c^i(\alpha_o, \eta_o), \quad i = 1,2,3,4,5,6 \tag{2.10.7}$$

which define the polarization of plane waves.

According to (2.9.20), (2.10.4) and (2.10.7), the plane waves are defined as a combination of

$$\psi_i = \psi_o^i e^{-j\eta_o x - j\alpha_o z - jk_y^i y} \quad \mathrm{i} = 1,2,..,6 \tag{2.10.8}$$

where $\psi_o^i = |\psi_c^i, \psi_d^i|^t$. The components ψ_d^i, $i = 1,2,3,4,5,6$ of order 3 are defined for the medium by equations (2.9.24) that relate ψ_d to ψ_c, considering $D_x = -j\eta_o$, $D_z = -j\alpha_o$.

As in electromagnetic scattering of plane waves, we propose a general theory of reflection of elastic plane waves using network modeling (Figure 2.10.1).

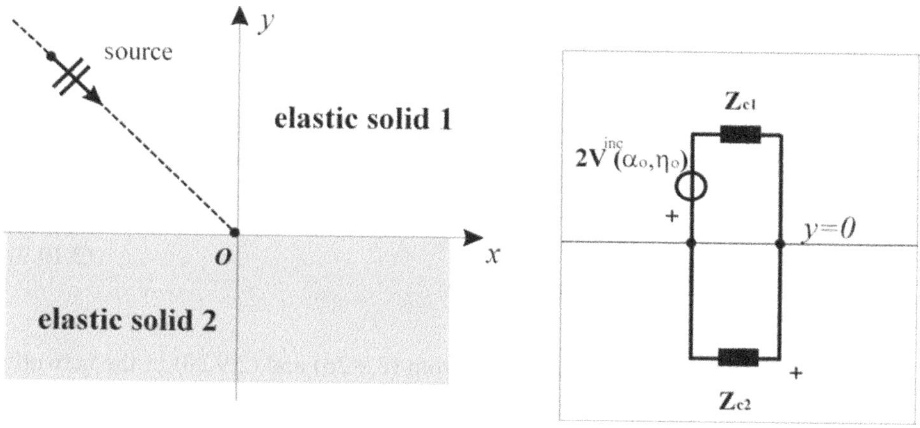

Figure 2.10.1. *Left: plane wave reflection between semi-infinite homogeneous elastic solids. Right: network modeling of the scattering problem in the spectral domain for a plane wave*

The network equivalent model on the right-hand side of Figure 2.10.1 is obtained by exploiting the network equivalent model in electromagnetics and circuit considerations (Felsen and Marcuvitz 1973; Daniele and Zich 2014).

First, we consider the equivalence between elasticity and electromagnetism at perfect boundary conditions. According to (2.2.18) and (2.9.26), in perfectly rigid elastic solids, the three components of velocity are zero, as in electromagnetism where the tangential components of magnetic fields are zero in a perfectly magnetically conducting (PMC) surface; at the same time, in free space, the continuous components of stress tensor are zero, as in electromagnetism where the tangential components of electric fields are zero in a PEC surface.

Moreover, since we use circuit equivalence for both electromagnetism and elasticity (section 2.9, in particular (2.9.40) and (2.9.42)), we apply the Thevenin theorem to the scattering of plane waves by an infinite layer.

Concerning the sub-region y>0, we obtain the Thevenin model by evaluating the open circuit voltage and the impedance in the absence of sources.

In this case, the impedance Z_{c1} is (2.9.40) while the open circuit voltage is obtained by imposing PMC termination at y=0. For this reason, the open circuit voltage is two times the voltage of the incident field due to the reflection coefficient equal to +1.

For y<0, the Thevenin model results in this being constituted of just the impedance Z_{c2} (2.9.42), due to the absence of sources.

With respect to Figure 2.10.1, we assume that the source consists of a plane wave incident from medium 1 (in accordance with the 6 waves defined in (2.10.8)):

$$\psi^{inc} = \psi_o e^{-j\eta_o x - j\alpha_o z - jk_y^{inc} y} = \begin{vmatrix} \psi_{co}^{inc} \\ \psi_{do}^{inc} \end{vmatrix} e^{-j\eta_o x - j\alpha_o z - jk_y^{inc} y} \quad (2.10.9)$$

where $jk_y^i = \gamma_i(\alpha_o, \eta_o)$ with i=inc=4,5,6. From (2.9.26) and (2.9.20) in the network model of Figure 2.10.1, we have

$$\mathbf{v}^{inc}(\alpha_o, \eta_o) = \begin{vmatrix} T_{2co}^{inc} \\ T_{4co}^{inc} \\ T_{6co}^{inc} \end{vmatrix} e^{-j\eta_o x - j\alpha_o z} \text{ at } y=0 \quad (2.10.10)$$

where $\begin{vmatrix} T_{2co}^{inc} \\ T_{4co}^{inc} \\ T_{6co}^{inc} \end{vmatrix}$ is the stress of the incident plane wave evaluated in x=0,y=0,z=0.

While considering *isotropic* homogeneous planar regions, the dispersion equation (2.10.5) considerably simplifies. In fact, in this case, we still have three non-null components of the velocity vector (z,x,y), but the six eigenvalues have the following form: longitudinal $\pm\gamma_P$ (P-wave with high phase velocity), transverse (shear) vertical $\pm\gamma_{SV}$ (SV wave) and transverse (shear) horizontal $\pm\gamma_{SH}$ (SH wave).

From (2.10.5), it yields $\gamma_P = \sqrt{\eta_o^2 + \alpha_o^2 - k_P^2}$, $\gamma_{SV} = \sqrt{\eta_o^2 + \alpha_o^2 - k_{SV}^2}$, $\gamma_{SH} = \sqrt{\eta_o^2 + \alpha_o^2 - k_{SH}^2}$ with $k_P = \omega\sqrt{\dfrac{\rho}{\lambda + 2\mu}}$, $k_{SH} = k_{SV} = \omega\sqrt{\dfrac{\rho}{\mu}}$, thus $\gamma_{SV} = \gamma_{SH}$.

Interesting properties can be observed in *two-dimensional* problems, i.e. for $D_z = \partial/\partial z = -j\alpha_o = 0$.

In this case, considering the half-space y > 0 in Figure 2.10.1 as an open waveguide filled by free space with the longitudinal direction y, the P-wave is polarized such that the velocity v is parallel to the direction of propagation, i.e. the P-wave is the longitudinal wave. The SV and SH waves instead have components that are transversal because v has components orthogonal to the direction of propagation. In particular, while for the SV wave both the components v_x and v_y are non-null, the SH plane presents only the non-null component v_z.

Taking into account the transverse field equations (2.9.23) and (2.9.24), we obtain that, for two-dimensional problems, the SH waves are independent/decoupled from P and SV waves, while P and SV are related together (Oliner 1969). This means if we illuminate a structure in the two-dimensional problem with an SH wave, only SH waves are excited. On the contrary, if we illuminate a structure in the two-dimensional problem with a P or SV wave, only P and SV waves are excited.

From (2.9.23) and (2.9.24), we also observe that SH polarization has only non-null ψ_c components v_z and T_4, while P and SV polarizations have combined P and SV non-null ψ_c components v_x, v_y, T_2 and T_6.

For the seismologist, V means vertical (i.e. component in the plane of incidence) and H means horizontal (i.e. component orthogonal to the plane of incidence) (see Figure 2.10.1).

A further simplification arises in the presence of the planar problem, where the two-dimensional problem is further simplified by assuming v_z=0 while the wave propagates on the x-y plane. The assumption of the planar problem is equivalent to making the SH polarization null in the two-dimensional problem. On the contrary, the anti-planar problem is equivalent to making only the SH polarization non-null in the two-dimensional problem.

If we consider a homogeneous isotropic two-dimensional elastic solid planar problem, we have T_4=0, T_5=0 from (2.9.23) and (2.9.24), where only P and SV waves are present. In this case, the wave vector ψ_c (2.9.21) is of order 4 instead of 6 and the vectors **V** and **I** are of order 2:

$$\psi_c = \left|T_2, T_6, v_1, v_2\right|^t \Rightarrow \quad \mathbf{V} = \begin{vmatrix} T_2 \\ T_6 \end{vmatrix} = \begin{vmatrix} T_{yy} \\ T_{yx} \end{vmatrix}, \quad \mathbf{I} = \begin{vmatrix} v_1 \\ v_2 \end{vmatrix} = \begin{vmatrix} v_x \\ v_y \end{vmatrix} \tag{2.10.11}$$

and $\mathbf{M}(\alpha_o, \eta_o)$ of (2.10.3), (2.10.5), (2.9.25) is explicitly

$$\mathbf{M}(\alpha_o = 0, \eta_o) = \begin{vmatrix} 0 & Dx & 0 & -j\rho\omega \\ \dfrac{Dx\lambda}{\lambda+2\mu} & 0 & \dfrac{-j(4Dx^2\lambda\mu + 4Dx^2\mu^2 + \lambda\rho\omega^2 + 2\mu\rho\omega^2)}{(\lambda+2\mu)\omega} & 0 \\ 0 & \dfrac{-j\omega}{\mu} & 0 & Dx \\ \dfrac{-j\omega}{\lambda+2\mu} & 0 & \dfrac{Dx\lambda}{\lambda+2\mu} & 0 \end{vmatrix} \quad (2.10.12)$$

with $D_x = -j\eta_o$ for the specific plane wave assumption (2.10.9).

The ordering of **V** and **I** components in (2.10.11) is important in order to obtain the correct interpretation in terms of power flow using the Poynting vector.

According to (2.10.11), we note that the SH plane wave is not present in this simplified problem as in the electromagnetic problem with parallel polarization where the Ez component is not present.

In fact, the eigenvalues of (2.10.3) with (2.10.12) are

$$\gamma_1 = \gamma_P = \sqrt{\eta_o^2 - \frac{\rho\omega^2}{\lambda+2\mu}} = \sqrt{\eta_o^2 - k_P} = -\gamma_3 \quad (2.10.13)$$

$$\gamma_2 = \gamma_{SV} = \sqrt{\eta_o^2 - \frac{\rho\omega^2}{\mu}} = \sqrt{\eta_o^2 - k_{SV}} = -\gamma_4 \quad (2.10.14)$$

and the corresponding phase velocities are

$$v_P = \frac{\omega}{k_P} = \sqrt{\frac{\lambda+2\mu}{\rho}} \quad (2.10.15)$$

$$v_{SV} = \frac{\omega}{k_{SV}} = \sqrt{\frac{\mu}{\rho}} \quad (2.10.16)$$

The eigenvectors of (2.10.3) with (2.10.12) are

$$\psi_c^1 = \begin{pmatrix} \dfrac{-j\sqrt{\lambda+2\mu}(2D_x^2\mu+\rho\omega^2)}{\omega\sqrt{-D_x^2(\lambda+2\mu)-\rho\omega^2}} \\ -\dfrac{2jD_x\mu}{\omega} \\ \dfrac{-D_x\sqrt{\lambda+2\mu}}{\sqrt{-D_x^2(\lambda+2\mu)-\rho\omega^2}} \\ 1 \end{pmatrix}; \psi_c^2 = \begin{pmatrix} \dfrac{2j\sqrt{\mu}\sqrt{-Dx^2\mu-\rho\omega^2}}{\omega} \\ \dfrac{j(-2D_x^2\mu-\rho\omega^2)}{D_x\omega} \\ \dfrac{\sqrt{-D_x^2\mu-\rho\omega^2}}{D_x\sqrt{\mu}} \\ 1 \end{pmatrix};$$

$$\psi_c^3 = \begin{pmatrix} \dfrac{j\sqrt{\lambda+2\mu}(2D_x^2\mu+\rho\omega^2)}{\omega\sqrt{-D_x^2(\lambda+2\mu)-\rho\omega^2}} \\ -\dfrac{2jD_x\mu}{\omega} \\ \dfrac{D_x\sqrt{\lambda+2\mu}}{\sqrt{-D_x^2(\lambda+2\mu)-\rho\omega^2}} \\ 1 \end{pmatrix} = \begin{pmatrix} -\psi_{c,1}^1 \\ \psi_{c,2}^1 \\ -\psi_{c,3}^1 \\ 1 \end{pmatrix};$$

$$\psi_c^4 = \begin{pmatrix} \dfrac{-2j\sqrt{\mu}\sqrt{-Dx^2\mu-\rho\omega^2}}{\omega} \\ \dfrac{j(-2Dx^2\mu-\rho\omega^2)}{Dx\omega} \\ -\dfrac{\sqrt{-Dx^2\mu-\rho\omega^2}}{Dx\sqrt{\mu}} \\ 1 \end{pmatrix} = \begin{pmatrix} -\psi_{c,1}^2 \\ \psi_{c,2}^2 \\ -\psi_{c,3}^2 \\ 1 \end{pmatrix}; \qquad (2.10.17)$$

where $D_x = -j\eta_o$ for the specific plane wave assumption (2.10.9).

A dimensional control always assures that the first and second components of these vectors have the same dimension and, similarly, the third and fourth components have dimensions that are those of the first and second components multiplied by an elastic admittance.

Henceforth, using (2.10.17), we formalize the homogeneous isotropic elastic solid planar problem in four dimensions instead of six. According to (2.10.13) and (2.10.14), ψ_c^1, ψ_c^2 are progressive waves, while ψ_c^3, ψ_c^4 are regressive. Moreover, $\psi_c^{(i)}$ can be interpreted as constants as follows:

$$\psi_c^i = \left| V_1^i, V_2^i, I_1^i, I_2^i \right|^t = \left| T_2^i, T_6^i, v_1^i, v_2^i \right|^t \tag{2.10.18}$$

2.11. Diffraction by a half-infinite crack in a homogeneous elastic solid planar problem

Let us now consider a crack in an arbitrary homogeneous elastic solid (Figure 2.11.1).

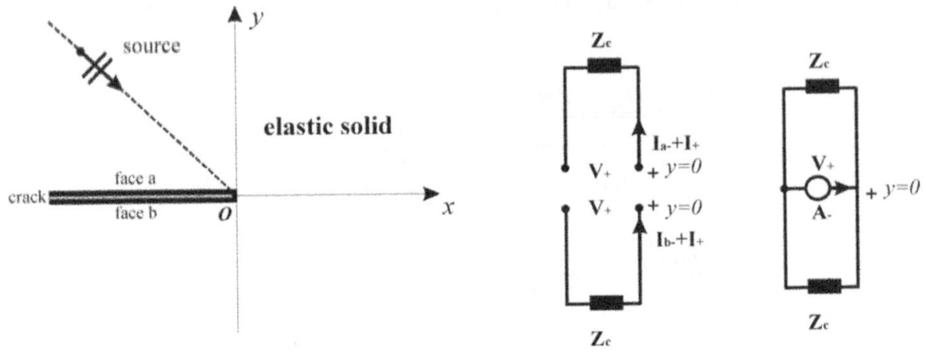

Figure 2.11.1. *Left: half-infinite crack in a homogeneous elastic solid planar problem illuminated by plane waves. Center, right: equivalent network models in the Laplace domain*

We define the following minus and plus spectra:

$$\mathbf{I}_{a-} = \int_{-\infty}^{0} \begin{vmatrix} v_1(\alpha_o, 0_+, x) \\ v_2(\alpha_o, 0_+, x) \\ v_3(\alpha_o, 0_+, x) \end{vmatrix} e^{j\eta x} dx, \ \mathbf{V}_{a-} = \int_{-\infty}^{0} \begin{vmatrix} T_2(\alpha_o, 0_+, x) \\ T_4(\alpha_o, 0_+, x) \\ T_6(\alpha_o, 0_+, x) \end{vmatrix} e^{j\eta x} dx \tag{2.11.1}$$

$$\mathbf{I}_{b-} = \int_{-\infty}^{0} \begin{vmatrix} v_1(\alpha_o, 0_-, x) \\ v_2(\alpha_o, 0_-, x) \\ v_3(\alpha_o, 0_-, x) \end{vmatrix} e^{j\eta x} dx, \ \mathbf{V}_{b-} = \int_{-\infty}^{0} \begin{vmatrix} T_2(\alpha_o, 0_-, x) \\ T_4(\alpha_o, 0_-, x) \\ T_6(\alpha_o, 0_-, x) \end{vmatrix} e^{j\eta x} dx \tag{2.11.2}$$

$$\mathbf{I}_{+} = \int_{0}^{\infty} \begin{vmatrix} v_1(\alpha_o, 0_+, x) \\ v_2(\alpha_o, 0_+, x) \\ v_3(\alpha_o, 0_+, x) \end{vmatrix} e^{j\eta x} dx, \ \mathbf{V}_{+} = \int_{0}^{\infty} \begin{vmatrix} T_2(\alpha_o, 0, x) \\ T_4(\alpha_o, 0, x) \\ T_6(\alpha_o, 0, x) \end{vmatrix} e^{j\eta x} dx \tag{2.11.3}$$

where (2.11.1) provides the velocity and stress spectra on the face a of the crack ($y=0_+$, $x<0$), (2.11.2) provides the velocity and stress spectra on the face b of the crack ($y=0_-$, $x<0$) and (2.11.3) provides the continuous velocity and stress spectra on the aperture ($y=0$, $x>0$).

Taking into account the boundary conditions of the crack, i.e. the null even component of the stress vectors T_2,T_4 and T_6 for $x<0, y=0_{+,-}$, we have it that the "voltage"

$$\mathbf{V}_{a-} = 0\,,\ \mathbf{V}_{b-} = 0 \tag{2.11.4}$$

This property recalls the behavior of the PEC half-plane in electromagnetic scattering.

As described in section 2.9 in (2.9.40)–(2.9.43) and considering bilateral Laplace transforms, we have

$$\mathbf{I}_{a-} + \mathbf{I}_+ = \vec{\mathbf{Y}}_c(\mathbf{V}_{a-} + \mathbf{V}_+) = \vec{\mathbf{Y}}_c\mathbf{V}_+ \tag{2.11.5}$$

$$\mathbf{I}_{b-} + \mathbf{I}_+ = -\overleftarrow{\mathbf{Y}}_c(\mathbf{V}_{b-} + \mathbf{V}_+) = -\overleftarrow{\mathbf{Y}}_c\mathbf{V}_+ \tag{2.11.6}$$

where $\vec{\mathbf{Y}}_c = \vec{\mathbf{Z}}_c^{-1}$, $\overleftarrow{\mathbf{Y}}_c = \overleftarrow{\mathbf{Z}}_c^{-1}$.

With reference to Figure 2.11.1 (left and center), (2.11.5) is a homogeneous equation due to the application of bilateral Laplace transforms, as discussed in the electromagnetic problem in section 2.3 (see section 1.4 for the absence of the source term).

By subtracting (2.11.5) from (2.11.6), we obtain the WH equation of the crack problem:

$$\mathbf{A}_- = \mathbf{Y}\mathbf{V}_+ \tag{2.11.7}$$

where the kernel $\mathbf{Y}$ is given by

$$\mathbf{Y} = \vec{\mathbf{Y}}_c + \overleftarrow{\mathbf{Y}}_c \tag{2.11.8}$$

and the quantity $\mathbf{A}_- = \mathbf{I}_{a-} - \mathbf{I}_{b-}$ is a minus function (see Figure 2.11.1 (right)).

Assuming plane wave illumination defined by

$$\psi^{inc} = \psi_o e^{-j\eta_o x - j\alpha_o z - jk_y^{inc} y} = \begin{vmatrix} \psi_{co}^{inc} \\ \psi_{do}^{inc} \end{vmatrix} e^{-j\eta_o x - j\alpha_o z - jk_y^{inc} y} \tag{2.11.9}$$

the solution of (2.11.7) is given by (1.5.13):

$$\mathbf{V}_+(\eta) = [\mathbf{Y}_+(\eta)]^{-1} \left[\mathbf{Y}_+(\eta_o)\mathbf{V}_+^{n.s.}(\eta) + \mathbf{Y}_-^{-1}(\eta_o)\mathbf{A}_-^{n.s.}(\eta) \right] \tag{2.11.10}$$

where $\mathbf{V}_+^{n.s.}(\eta)$ and $\mathbf{A}_-^{n.s.}(\eta)$ are obtained through the GO analysis.

The problem is reduced to the factorization of the matrix $\mathbf{Y}(\eta) = \mathbf{Y}_-(\eta)\mathbf{Y}_+(\eta)$.

While, in very simple cases, the factorization of $\mathbf{Y}(\eta)$ can be obtained in the closed form, in general we can use the Fredholm factorization method.

Equation (2.11.10) provides the spectra of the stress $\mathbf{V}(\eta,0) = \mathbf{V}_-(\eta) + \mathbf{V}_+(\eta) = \mathbf{V}_+(\eta)$ in the plane y=0. To evaluate the spectra for each value of y, we can use the first of equations (2.9.35) and (2.9.36), which can be rewritten in the form:

$$\begin{aligned} \mathbf{V}(\eta,y) &= C_1(\eta)e^{-\gamma_1(\eta)y}\mathbf{V}^{(1)}(\eta) + C_2(\eta)e^{-\gamma_2(\eta)y}\mathbf{V}^{(2)}(\eta) \\ &\quad + C_3(\eta)e^{-\gamma_3(\eta)y}\mathbf{V}^{(3)}(\eta) \qquad y>0 \\ \mathbf{V}(\eta,y) &= C_4(\eta)e^{-\gamma_4(\eta)y}\mathbf{V}^{(4)}(\eta) + C_5(\eta)e^{-\gamma_5(\eta)y}\mathbf{V}^{(5)}(\eta) \\ &\quad + C_6(\eta)e^{-\gamma_6(\eta)y}\mathbf{V}^{(6)}(\eta) \qquad y<0 \end{aligned} \tag{2.11.11}$$

The scalars $C_1(\eta)$, $C_2(\eta)$, $C_3(\eta)$ that provide the progressive spectra can be evaluated by forcing y=0 that yields the three linear equations:

$$\mathbf{V}(\eta,0) = \mathbf{V}_+(\eta) = C_1(\eta)\mathbf{V}^{(1)}(\eta) + C_2(\eta)\mathbf{V}^{(2)}(\eta) + C_3(\eta)\mathbf{V}^{(3)}(\eta) \tag{2.11.12}$$

Similarly, to obtain $C_4(\eta)$, $C_5(\eta)$, $C_6(\eta)$ that provide the regressive spectra, we can solve the three linear equations

$$\mathbf{V}_+(\eta) = C_4(\eta)\mathbf{V}^{(4)}(\eta) + C_5(\eta)\mathbf{V}^{(5)}(\eta) + C_6(\eta)\mathbf{V}^{(6)}(\eta) \tag{2.11.13}$$

To obtain the diffracted stress component from the spectra, we use the inverse transform (2.7.7):

$$\begin{vmatrix} T_2(x,y) \\ T_4(x,y) \\ T_6(x,y) \end{vmatrix} = \frac{1}{2\pi}\int_{B_r} \mathbf{V}(\eta, y) e^{-j\eta x} dx \qquad (2.11.14)$$

where B_r is the Bromwich contour. Far-field evaluations can be obtained using the saddle point method.

2.12. Diffraction by a half-infinite crack in a homogeneous isotropic two-dimensional elastic solid problem

We consider the two-dimensional ($\alpha_o = 0$) diffraction problem of a half-infinite crack located in a homogeneous isotropic elastic solid. This problem can be decoupled into planar (P and SV waves) and anti-planar (SH waves) problems, as discussed and defined in section 2.10 and in (Oliner 1969) and (Achenbach 1973).

Without loss of generality, we first assume that the incident plane wave with the direction φ_o is an SH wave with the wavelength $k_{SH} = \omega\sqrt{\frac{\rho}{\mu}}$ (anti-planar problem). In this case, no coupling is present between SH and P/SV waves. From (2.9.23) and (2.9.24), only the velocity component v_3 and the stress components $T_4 = T_{yz} = \frac{\mu}{j\omega}\frac{\partial v_3}{\partial y}$ and $T_5 = T_{xz} = \frac{\mu}{j\omega}\frac{\partial v_3}{\partial x}$ are non-null.

On the half-infinite crack, the boundary condition is $T_4 = 0$.

In this anti-planar case, the study of the diffraction by the half-infinite crack is straightforward, since we use the electromagnetic analogy consisting of the scattering problem of an Hz-polarized plane wave at normal incidence (similar to the procedure presented in sections 2.3–2.5, but with different polarizations): $v_3(x,y)$ is $H_z(x,y)$, $T_4(x,y)$ is $E_x(x,y)$, μ is $1/\varepsilon_o$ and ρ is μ_o.

To deduce the WH equation of the problem, we apply the circuit interpretation of transmission line equations, as shown in Figure 2.11.1.

In this case, (2.11.1)–(2.11.3) reduce to

$$\mathbf{I}_{a-} = \int_{-\infty}^{0} v_3(\alpha_o, 0_+, x) e^{j\eta x} dx, \ \mathbf{V}_{a-} = \int_{-\infty}^{0} T_4(\alpha_o, 0_+, x) e^{j\eta x} dx = 0 \qquad (2.12.1)$$

$$\mathbf{I}_{b-} = \int_{-\infty}^{0} v_3(\alpha_o, 0_-, x) e^{j\eta x} dx, \ \mathbf{V}_{b-} = \int_{-\infty}^{0} T_4(\alpha_o, 0_-, x) e^{j\eta x} dx = 0$$

$$\mathbf{I}_{+} = \int_{0}^{\infty} v_3(\alpha_o, 0_+, x) e^{j\eta x} dx, \ \mathbf{V}_{+} = \int_{0}^{\infty} T_4(\alpha_o, 0, x) e^{j\eta x} dx$$

with $\alpha_o = 0$.

Considering $T_4 = T_{yz} = \dfrac{\mu}{j\omega} \dfrac{\partial v_3}{\partial y}$ from (2.9.23) and transmission line modeling along y>0 with propagation $e^{-\gamma_{SH}(\eta) y}$ with $\gamma_{SH} = \sqrt{\eta^2 - k_{SH}^2}$ (section 2.10), we obtain, in the Laplace domain, the scalar characteristic admittance for the anti-planar problem:

$$\vec{Y}_c = \overleftarrow{Y}_c = \frac{Y}{2} = \frac{\omega}{\mu\sqrt{k_{SH}^2 - \eta^2}}$$

As expected, we deal with the WH equation (2.11.7) that is similar to the electromagnetics (2.3.7), but with different polarizations. The factorization of Y is immediate:

$$Y = \frac{2\omega}{\mu\sqrt{k_{SH}^2 - \eta^2}} = \frac{\sqrt{2\omega/\mu}}{\sqrt{k_{SH} + \eta}} \frac{\sqrt{2\omega/\mu}}{\sqrt{k_{SH} - \eta}}$$

Assuming a source $v_3^{inc}(x,0) = v_o e^{-j\eta_o x}$ with $\eta_o = -k_{SH}\cos\varphi_o$, we obtain the solutions (2.11.10), i.e. in explicit form:

$$V_+(\eta) = -\frac{\mu\sqrt{k_{SH} - \eta}\sqrt{k_{SH} + \eta_o}}{\omega} \frac{j v_o}{\eta - \eta_o} \qquad (2.12.2)$$

This is analogous to the solution of electromagnetic scattering of a PEC half-plane by an Hz-polarized plane wave at normal incidence, whose solution is (2.12.2) with $v_o = H_o$, where H_o is the intensity of the Hz wave and $k_{SH} = k$ of the medium.

Equation (2.12.2) represents the spectral solution of the anti-planar two-dimensional diffraction problem by a half-infinite crack located in a homogeneous isotropic elastic solid.

Conversely, in the diffraction planar two-dimensional problem by a half-infinite crack located in a homogeneous isotropic elastic solid, only P and/or SV waves are excited by incident P and/or SV waves (see sections 2.10 and 2.11). In particular, no SH wave is present; therefore, $T_4 = 0$ and $v_3 = 0$ everywhere (section 2.10).

In this problem, the matrix $\mathbf{Y} = \vec{\mathbf{Y}} + \overleftarrow{\mathbf{Y}}$ (2.11.8) is of order 2, and here we anticipate that $\mathbf{Y}$ can be factorized in the closed form.

In accordance with (2.9.40)–(2.9.43) and the description of the planar problem, we have:

$$\vec{\mathbf{Z}}_c = \begin{vmatrix} V_1^{(1)} & V_1^{(2)} \\ V_2^{(1)} & V_2^{(2)} \end{vmatrix} \cdot \begin{vmatrix} I_1^{(1)} & I_1^{(2)} \\ I_2^{(1)} & I_2^{(2)} \end{vmatrix}^{-1}, \quad \overleftarrow{\mathbf{Z}}_c = \begin{vmatrix} V_1^{(3)} & V_1^{(4)} \\ V_2^{(3)} & V_2^{(4)} \end{vmatrix} \cdot \begin{vmatrix} I_1^{(3)} & I_1^{(4)} \\ I_2^{(3)} & I_2^{(4)} \end{vmatrix}^{-1} \tag{2.12.3}$$

where $\mathbf{V}^{(i)} = \begin{vmatrix} V_1^{(i)} \\ V_2^{(i)} \end{vmatrix}$ and $\mathbf{I}^{(i)} = \begin{vmatrix} I_1^{(i)} \\ I_2^{(i)} \end{vmatrix}$ (i=1,2,3,4) are, respectively, the first two rows and the last two rows of the eigenvectors defined by ψ_c^i (i=1,2,3,4) in (2.10.17) assuming $Dx = -j\eta$.

By recalling that the notation $\psi_{c,j}^{(i)}$ is the j component of the i eigenvector, after some mathematical manipulation of

$$\vec{\mathbf{Y}}_c = \begin{vmatrix} \psi_{c,1}^{(1)} & \psi_{c,1}^{(2)} \\ \psi_{c,2}^{(1)} & \psi_{c,2}^{(2)} \end{vmatrix}^{-1} \cdot \begin{vmatrix} \psi_{c,3}^{(1)} & \psi_{c,3}^{(2)} \\ \psi_{c,4}^{(1)} & \psi_{c,4}^{(2)} \end{vmatrix}; \quad \overleftarrow{\mathbf{Y}}_c = \begin{vmatrix} \psi_{c,1}^{(3)} & \psi_{c,1}^{(4)} \\ \psi_{c,2}^{(3)} & \psi_{c,2}^{(4)} \end{vmatrix}^{-1} \cdot \begin{vmatrix} \psi_{c,3}^{(3)} & \psi_{c,3}^{(4)} \\ \psi_{c,4}^{(3)} & \psi_{c,4}^{(4)} \end{vmatrix} \tag{2.12.4}$$

we obtain

$$\vec{\mathbf{Y}}_c = \begin{vmatrix} n & b_1/2 \\ b_2/2 & -n \end{vmatrix}, \quad \overleftarrow{\mathbf{Y}}_c = \begin{vmatrix} -n & b_1/2 \\ b_2/2 & n \end{vmatrix} \tag{2.12.5}$$

with

$$n = \frac{\eta\omega(-2\eta^2\mu\sqrt{\lambda+2\mu} + \sqrt{\lambda+2\mu}\rho\omega^2 + 2\sqrt{\mu}\sqrt{\eta^2\mu-\rho\omega^2}\sqrt{\eta^2(\lambda+2\mu)-\rho\omega^2})}{-4\eta^4\mu^2\sqrt{\lambda+2\mu} - \sqrt{\lambda+2\mu}\rho^2\omega^4 + 4\eta^2(\mu\sqrt{\lambda+2\mu}\rho\omega^2 + \mu^{3/2}\sqrt{\eta^2\mu-\rho\omega^2}\sqrt{\eta^2(\lambda+2\mu)-\rho\omega^2})} \tag{2.12.6}$$

$$b_1 = \frac{i\sqrt{\lambda+2\mu}\,\rho\omega^3\sqrt{\eta^2\mu-\rho\omega^2}}{\sqrt{\mu}\left(4\eta^4\mu^2\sqrt{\lambda+2\mu}+\sqrt{\lambda+2\mu}\,\rho^2\omega^4-4\eta^2\left(\mu\sqrt{\lambda+2\mu}\,\rho\omega^2+\mu^{3/2}\sqrt{\eta^2\mu-\rho\omega^2}\sqrt{\eta^2(\lambda+2\mu)-\rho\omega^2}\right)\right)}$$

(2.12.7)

$$b_2 = \frac{i\rho\omega^3\sqrt{\eta^2(\lambda+2\mu)-\rho\omega^2}}{4\eta^4\mu^2\sqrt{\lambda+2\mu}+\sqrt{\lambda+2\mu}\,\rho^2\omega^4-4\eta^2\left(\mu\sqrt{\lambda+2\mu}\,\rho\omega^2+\mu^{3/2}\sqrt{\eta^2\mu-\rho\omega^2}\sqrt{\eta^2(\lambda+2\mu)-\rho\omega^2}\right)}$$

(2.12.8)

The anti-diagonal form of (2.12.5) substituted in (2.11.7)–(2.11.8) allows us to obtain the following uncoupled scalar WH equations:

$$I_{a1-} - I_{b1-} = b_1 V_{2+} \tag{2.12.9}$$

$$I_{a2-} - I_{b2-} = b_2 V_{1+} \tag{2.12.10}$$

We observe that (2.12.10) can be directly obtained by considering the half-plane problem where the velocity component v_x is null on the crack and the stress T_{yx} is null on the aperture. Similarly, (2.12.9) can be directly obtained by considering the half-plane problem where the velocity component v_y is null on the crack and the stress T_{yy} is null on the aperture. In the literature, the above two problems are, respectively, called symmetric and antisymmetric problems (Achenbach 1973).

Using the matrix notation, we can rewrite the matrix kernel $\mathbf{Y} = \vec{\mathbf{Y}} + \bar{\mathbf{Y}}$ in the factorized form

$$\mathbf{Y} = \begin{vmatrix} 0 & b_1 \\ b_2 & 0 \end{vmatrix} = \begin{vmatrix} b_1 & 0 \\ 0 & b_2 \end{vmatrix} . \sigma = \begin{vmatrix} b_{1-} & 0 \\ 0 & b_{2-} \end{vmatrix} \begin{vmatrix} b_{1+} & 0 \\ 0 & b_{2+} \end{vmatrix} \sigma \tag{2.12.11}$$

where the constant matrix σ is defined by

$$\sigma = \begin{vmatrix} 0 & 1 \\ 1 & 0 \end{vmatrix}, \quad \sigma^2 = \begin{vmatrix} 1 & 0 \\ 0 & 1 \end{vmatrix} \tag{2.12.12}$$

According to (2.12.5)–(2.12.8), we observe that

$$b_2(\eta) = -\frac{2j\omega k_{SV}^2}{\mu}\frac{\sqrt{\eta^2 - k_P^2}}{R(\eta)} \tag{2.12.13}$$

$$b_1(\eta) = -\frac{2j\omega k_{SV}^2}{\mu} \frac{\sqrt{\eta^2 - k_{SV}^2}}{R(\eta)} \tag{2.12.14}$$

where $R(\eta)$ is the Rayleigh function defined by

$$R(\eta) = (2\eta^2 - k_{SV}^2)^2 - 4\eta^2 \sqrt{\eta^2 - k_{SV}^2} \sqrt{\eta^2 - k_P^2} \tag{2.12.15}$$

and the wavelength numbers k_{SV} and k_P are real positive and defined in (2.10.13)–(2.10.16) for the lossless case as assumed in (2.9.12)–(2.9.13).

According to (2.12.11)–(2.12.14), we observe that

$$b_{1\pm}(\eta) = \sqrt{\frac{\eta \mp k_{SV}}{\eta \mp k_P}} b_{2\pm}(\eta) \tag{2.12.16}$$

In order to obtain explicit expressions for (2.12.16), the factorization of the Rayleigh function is important:

$$R(\eta) = (2\eta^2 - k_{SV}^2)^2 - 4\eta^2 \sqrt{\eta^2 - k_{SV}^2} \sqrt{\eta^2 - k_P^2} \tag{2.12.17}$$

We observe that since $\mathrm{R}(\eta) = O[\eta^2]$ as $\eta \to \infty$, it yields $b_1(\eta) = O[\eta]$, $b_2(\eta) = O[\eta]$ as $\eta \to \infty$ and the Rayleigh function $\mathrm{R}(\eta)$ vanishes at $\eta = \pm k_R = \frac{\omega}{c_R}$, where c_R is the velocity of Rayleigh surface waves.

An exact expression of k_R as a function of k_P and k_{SV} ($k_{SV} > k_P$, in practice) is obtained after some mathematical manipulations and we report it here for literature reference:

$$k_{\mathrm{R}} = \frac{k_{SV}}{2\sqrt{3}} \sqrt{\frac{2^{1/3} f^{2/3} + 8 2^{2/3} k_P^4 + 6 f^{1/3} k_{SV}^2 + 6 2^{2/3} k_{SV}^4 - 4 k_P^2 (f^{1/3} + 3 2^{2/3} k_{SV}^2)}{f^{1/3}(-k_P^2 + k_{SV}^2)}} \tag{2.12.18}$$

with

$$f = -32k_P^6 + 99k_P^4k_{SV}^2 - 90k_P^2k_{SV}^4 + 27k_{SV}^6 + 3\sqrt{3}k_{SV}\sqrt{-64k_P^6 + 107k_P^4k_{SV}^2 - 62k_P^2k_{SV}^4 + 11k_{SV}^6}\,(k_{SV}^2 - k_P^2) \qquad (2.12.19)$$

In order to apply the logarithmic factorization to $b_1(\eta)$ and $b_2(\eta)$, we normalize the functions to obtain bounded values as $\eta \to \infty$ by defining

$$K(\eta) = \frac{\eta^2 - k_R^2}{R(\eta)} = K_-(\eta)K_+(\eta) \qquad (2.12.20)$$

The function $K(\eta)$ and its factorization have been studied in (Achenbach 1973). The explicit factorization of Y is obtained as follows:

$$\mathbf{Y} = \begin{vmatrix} 0 & b_1 \\ b_2 & 0 \end{vmatrix} = \begin{vmatrix} b_1 & 0 \\ 0 & b_2 \end{vmatrix} . \sigma = \begin{vmatrix} b_{1-} & 0 \\ 0 & b_{2-} \end{vmatrix} \begin{vmatrix} b_{1+} & 0 \\ 0 & b_{2+} \end{vmatrix} \sigma = Y_- Y_+ \qquad (2.12.21)$$

where:

$$b_{1+} = \frac{\sqrt{\eta - k_{SV}}}{\eta \ - k_R} K_+(\eta), \quad b_{1-} = \frac{\mathrm{b}}{\mathrm{b}_{1+}} \qquad (2.12.22)$$

$$b_{2+} = \frac{\sqrt{\eta - k_P}}{\eta \ - k_R} K_+(\eta), \quad b_{2-} = \frac{\mathrm{b}}{\mathrm{b}_{2+}} \qquad (2.12.23)$$

We recall that the WH equations have the form $\mathbf{A}_- = \mathbf{Y}\mathbf{V}_+$ (2.11.7). Assuming that the incident plane wave introduces a non-standard pole η_o for the plus function $\mathbf{V}_+(\eta)$, we rewrite the WH equation in the form:

$$Y_+(\eta)\mathbf{V}_+(\eta) - Y_+(\eta_o)\mathbf{V}_+^{n.s}(\eta) = Y_-^{-1}(\eta)\mathbf{A}_-(\eta) - Y_+(\eta_o)\mathbf{V}_+^{n.s}(\eta) = \mathbf{w} \qquad (2.12.24)$$

where $\mathbf{V}_+^{n.s}(\eta)$ is defined (2.10.10) as the Laplace transform:

$$\mathbf{V}_+^{n.s}(\eta) = \mathbf{V}_+^{G.O}(\eta) = \mathbf{V}_+^{inc}(\eta) = \frac{j}{\eta - \eta_o} \begin{vmatrix} T_{yy}^o \\ T_{yx}^o \end{vmatrix} \qquad (2.12.25)$$

where $\begin{vmatrix} T_{yy}^o \\ T_{yx}^o \end{vmatrix}$ is the stress of the incident plane wave evaluated at the origin (y=0).

Considering that

$$V_+(\eta) = O[1/\eta^{\nu}] \quad (\nu>0) \text{ as } \eta \to \infty \tag{2.12.26}$$

$$A_-(\eta) = O[1/\eta^{\nu+1}] \quad (\nu>0) \text{ as } \eta \to \infty \tag{2.12.27}$$

$$\mathbf{V}_+^{n.s}(\eta) = O(1/\eta) \text{ as } \eta \to \infty \tag{2.12.28}$$

the entire vector function **w** in (2.12.24) is null, which yields the solution:

$$\mathbf{V}_+(\eta) = Y_+^{-1}(\eta) Y_+(\eta_o) \mathbf{V}_+^{n.s}(\eta) \tag{2.12.29}$$

$$\mathbf{A}_-(\eta) = Y_-(\eta) Y_+(\eta_o) \mathbf{V}_+^{n.s}(\eta) \tag{2.12.30}$$

Equation (2.12.29) provides the spectral solution of the planar two-dimensional diffraction problem by a half-infinite crack located in a homogeneous isotropic elastic solid.

The asymptotic estimation of diffracted stress is obtained by a simplified form of (2.11.14):

$$\begin{vmatrix} T_{yy}(x,y) \\ T_{yx}(x,y) \end{vmatrix} = \frac{1}{2\pi} \int_{B_r} \mathbf{V}(\eta, y) e^{-j\eta x} dx \tag{2.12.31}$$

According to (2.10.13)–(2.10.14), we have

$$\gamma_1(\eta) = -j\xi_P(\eta),\ \gamma_2(\eta) = -j\xi_{SV}(\eta),\ \gamma_3(\eta) = j\xi_P(\eta),\ \gamma_4(\eta) = -j\xi_{SV}(\eta) \tag{2.11.32}$$

where:

$$\xi_P(\eta) = \sqrt{k_P^2 - \eta^2}\ ,\ \xi_{SV}(\eta) = \sqrt{k_{SV}^2 - \eta^2} \tag{2.12.33}$$

Consequently, we can rewrite the field as

$$\mathbf{V}(\eta, y) = \mathbf{V}_P(\eta, y) + \mathbf{V}_{SV}(\eta, y) \tag{2.12.34}$$

where the progressive components for y>0 are

$$\begin{aligned}\mathbf{V}_P(\eta,y) &= C_1(\eta)\mathbf{V}_1(\eta)e^{-j\xi_P(\eta)y}\\ \mathbf{V}_{SV}(\eta,y) &= C_2(\eta)\mathbf{V}_2(\eta)e^{-j\xi_{SV}(\eta)y}\end{aligned} \tag{2.12.35}$$

and the regressive components for y<0 are

$$\begin{aligned}\mathbf{V}_P(\eta,y) &= C_3(\eta)\mathbf{V}_3(\eta)e^{j\xi_P(\eta)y}\\ \mathbf{V}_{SV}(\eta,y) &= C_4(\eta)\mathbf{V}_4(\eta)e^{j\xi_{SV}(\eta)y}\end{aligned} \tag{2.12.36}$$

It yields the contributions

$$\begin{vmatrix}T_{yy}(x,y)\\ T_{yx}(x,y)\end{vmatrix} = \begin{vmatrix}T_{yy}^{P}(x,y)\\ T_{yx}^{P}(x,y)\end{vmatrix} + \begin{vmatrix}T_{yy}^{SV}(x,y)\\ T_{yx}^{SV}(x,y)\end{vmatrix} \tag{2.12.37}$$

where the progressive field (y>0) is given by

$$\begin{vmatrix}T_{yy}^{P}(x,y)\\ T_{yx}^{P}(x,y)\end{vmatrix} = \frac{1}{2\pi}\int_{B_r} C_1(\eta)\mathbf{V}_1(\eta)e^{-j\eta x}e^{-j\xi_P(\eta)y}dx \tag{2.12.38}$$

$$\begin{vmatrix}T_{yy}^{SV}(x,y)\\ T_{yx}^{SV}(x,y)\end{vmatrix} = \frac{1}{2\pi}\int_{B_r} C_2(\eta)\mathbf{V}_2(\eta)e^{-j\eta x}e^{-j\xi_V(\eta)y}dx \tag{2.12.39}$$

and the regressive field (y<0) is given by

$$\begin{vmatrix}T_{yy}^{P}(x,y)\\ T_{yx}^{P}(x,y)\end{vmatrix} = \frac{1}{2\pi}\int_{B_r} C_3(\eta)\mathbf{V}_3(\eta)e^{-j\eta x}e^{j\xi_P(\eta)y}dx \tag{2.12.40}$$

$$\begin{vmatrix}T_{yy}^{SV}(x,y)\\ T_{yx}^{SV}(x,y)\end{vmatrix} = \frac{1}{2\pi}\int_{B_r} C_4(\eta)\mathbf{V}_4(\eta)e^{-j\eta x}e^{j\xi_V(\eta)y}dx \tag{2.12.41}$$

Since, in our example, we assume the spectrum of the plus function $\mathbf{V}_+(\eta)$ to be non-standard, the Bromwich contour must be assumed to be above the pole η_o since $\mathrm{Im}[\eta_o] > 0$.

Although the plus and minus functions are vectors, the procedure to asymptotically estimate the far fields remains identical to that of section 2.7 based on the SDP method, since (2.12.38)–(2.12.41) are of the same form of (2.7.15) with the same $q(\eta)$ function.

Alternatively, we recall that the diffraction by a half-infinite crack in a homogeneous isotropic elastic solid planar problem is a classical canonical problem of elasticity, which, in the literature, has been solved by decomposing it into two distinct problems: the symmetric problem and the antisymmetric problem (Achenbach 1973) and (Budaev 1995).

The symmetric problem is defined by the following boundary condition on the crack:

$$I_{a1-} = I_{b1-}, V_{a1-} = V_{b1-}, V_{a2-} = -V_{b2-}, I_{a2-} = -I_{b2-} \tag{2.12.42}$$

while the antisymmetric problem is defined by

$$I_{a1-} = -I_{b1-}, V_{a1-} = -V_{b1-}, V_{a2-} = +V_{b2-}, I_{a2-} = +I_{b2-} \tag{2.12.43}$$

The WH formulation is again based on equations (2.11.5) and (2.11.6).

Taking into account the boundary conditions (2.12.42) and (2.12.43), after some mathematical manipulations, it is possible to reduce the problem to two scalar WH equations for both the symmetric and asymmetric cases.

3

The Wiener-Hopf Theory for Angular Region Problems

3.1. A brief history of the classical methods for studying angular regions

The first rigorous studies of waves in the presence of geometric discontinuities are due to Poincare (1892) and Sommerfeld (1896), who considered the presence of a half-plane in free space. The generalization of the half-plane problem leads to the wedge problem. It constituted an important and challenging subject within applied mathematics in the last century. A large number of papers have dealt with the wedge problem in many disciplines such as electromagnetism, acoustics, hydrodynamics, fracture mechanics and so on. The wedge can be classified into impenetrable and penetrable. The impenetrable wedge results from the introduction of approximate boundary conditions on the surfaces of wedges. For example, in electromagnetics, the classical condition $E_{\tan} = 0$ at the surface of a metal approximates the behavior of the electrical field near a metallic wedge at microwave frequencies very well. This impenetrability condition considerably simplifies the study of wedge problems. In fact, the external problem (i.e. the evaluation of fields outside the wedge) is decoupled from the internal problem (i.e. the evaluation of fields inside the wedge). Considerable advances have been made to obtain more realistic approximate boundary conditions (see for example Senior and Volakis (1995)). However, although sophisticated boundary conditions are available, this approximation is not sufficient to deal with penetrable wedges, i.e. constituted by two communicating angular regions. From a theoretical point of view, it is important to classify wedge scattering into problems where only one wave number is involved and problems where more wave numbers are introduced. In electromagnetism, different wave

---

This book consists of two volumes; any references to Chapters 4, 5 or 6 can be found in the second volume: *Scattering and Diffraction by Wedges 2.*

numbers result from the presence of more than one isotropic medium or from the presence of anisotropic materials. For example, problems where impenetrable wedges are immersed in an isotropic medium belong to the first class. On the contrary, if the medium surrounding the impenetrable wedge is an anisotropic material, then we deal with the presence of more than one wave number. Another case of multiple wave numbers is the penetrable wedge where at least two isotropic media are present. While considering wave motion in elasticity, we can also have at least two wave numbers in the presence of isotropic media.

Several methods are available in the literature to study wedges where only one wave number is involved. According to the author's opinion, the most popular one is the Sommerfeld–Malyuzhinets (SM) technique (Senior and Volakis 1995; Malyuzhinets, 1958a,b; Budaev 1995; Babich *et al.* 2008; Bernard 1998). The importance of this method has been well documented by a large number of papers. The original SM method has greatly advanced over the course of time. The key point in the SM method is the Sommerfeld representation of longitudinal electromagnetic fields.

Other similar representations developed by others, although simple (Williams 1959) or ingenious (Peters 1950; 1952), are not so popular since they introduce procedures that do not have the same physical interpretability of the SM method.

In the framework of field representations to solve wedge problems, the Kontorovich–Lebedev (KL) transform method (Jones 1964; Osipov 1993) seems to be the more general. However, the KL transforms are also related to the Sommerfeld integral (Malyuzhinets, 1958b; Daniele, 2003b). Although Sommerfeld representations allow us to immediately use powerful techniques for obtaining the far fields (saddle point technique) and the near fields (Watson lemma), the solutions obtained using the KL transforms do not have these immediate characteristics.

The methods mentioned previously are specialized methods for wedge problems.

Alternatively, it is also possible to obtain solutions of wedge problems using classical general methods, such as the separation of variables method (MacDonald 1915) or Green's function for angular regions (Felsen and Marcuvitz 1973) However, for application to general wedge problems, the last techniques seem to be less powerful with respect to the SM method (Budaev 1995).

For wedge problems with multiple wave numbers, the literature offers a vast amount of studies. In particular, the Sommerfeld integral can also be extended to the study of isotropic wedges in elasticity.

The penetrable wedge problem (e.g. a dielectric wedge in electromagnetics) is a very difficult problem that has aroused the interest of many authors for a long time. However, the closed-form solution for this problem remains unknown, even for the simplest cases.

An interesting attempt to solve the dielectric wedge problem was proposed by Radlow (1964) who obtained the solution related to the diffraction by the right-angled dielectric wedge, but unfortunately, this solution was wrong (Kraut and Lehaman 1969). The reason for the interest in the Radlow method is that it introduces multidimensional Wiener-Hopf equations. However, the solution of these equations yields factorization problems that need function-theoretic techniques employing two complex variables. Working on two-dimensional complex planes is very difficult, as it can easily introduce errors, as demonstrated in Albani (2007), about the quarter-plane scattering problem.

Using classical methods, the best way to deal with the penetrable wedge problem is the reduction to integral equations (Budaev 1995; Rawlins 1977; Salem *et al.* 2006).

Considering the above, the SM method is undoubtedly the most important classical method to handle wedge problems, as demonstrated by a number of recent books (Babich *et al.* 2008; Bernard 1998; Lyalinov and Zhu 2013).

3.2. Introduction to the generalized Wiener-Hopf technique

The generalized Wiener-Hopf technique (GWHT) described in this book is a novel method that completes the spectral technique and allows us to study field problems in angular regions. For a more reliable mathematical formulation, according to these authors, the advantages of the Wiener-Hopf (WH) method over the SM method are evident when closed-form solutions are not possible and approximate techniques become necessary. Furthermore, while the SM method is limited to angular regions where wave equations of Helmholtz type hold, the GWHT provides a general method for formulating arbitrary linear wedge problems. However, despite the preference of these authors for the WH technique, they claim that the SM method remains fundamental for studying diffraction by wedge problems.

The SM method and the Wiener-Hopf technique are different but closely related methods. For example, Keller (1979) posed the following question: "What features of the methods account for this difference?" Furthermore, Keller notes that "it might be helpful to understand this in order to predict the success of other methods". We agree with this opinion expressed by the giant of diffraction. Moreover, we believe that the SM and WH methods applied to the same problems can determine a helpful synergy. In the past, the SM and WH methods were considered to be disconnected, in particular because the SM method was traditionally defined with the angular complex representation while the WH method was traditionally defined in the Laplace domain. We observe that the two methods have significant points of similarity when the representation of problems in both methods is expressed, in terms of difference equations in the complex angular domain. The two methods show their diversity in the solution procedures that are completely different and effective. Both similarity and diversity properties are of advantage, as discussed in Keller's article "Progress and prospects in the theory of linear wave propagation" (Keller 1979).

For example, when dealing with isotropic media, the aspects of analytic continuations of approximate solutions, the evaluation of the spectra for any direction of the angular region and the far-field evaluation of the integrals are topics that are very difficult to study in the framework of the WH technique.

Therefore, we often recall in this book important results obtained with the SM method using the principle that the best procedures are those that take advantage of the possible different representations and formulations available in the literature.

From a mathematical point of view, while the SM method reduces the wedge problems to difference equation solutions, the fundamental problem in the WH technique is the factorization of matrix kernels.

The tool used to solve the difference equations in the SM method is the Malyuzhinets–Fourier (MF) transform.

Conversely, the WH factorization problem can be solved using several techniques developed through many years of application of the WH method.

As is well known, the possibility of obtaining closed-form solutions is limited.

In practice, in the SM method, difference equations must be of order 1 or, in general, must have constant coefficients. Similarly, general formulas to factorize WH kernels are available for scalar kernels or, in general, for matrix kernels having particular forms similar to those considered in Chapter 1.

Many efforts have been accomplished to extend the possibility of solving vector difference equations (Bernard 1998; Borovikov 1998; Senior and Legault 1998; Senior 1999; Antipov and Silvestrov 2004; Antipov and Silvestrov 2004a; Antipov and Silvestrov 2006) or to factorize kernel matrices of arbitrary order (Daniele and Zich 2014), but until now, general methods are not available. According to the authors of this book, due to the more complete mathematical theory, the research on the general factorization technique is more advanced with respect to the solution of general difference equations.

In any case, the best way to obtain efficient and general approximate solutions is to reduce the wave scattering problems to Fredholm integral equations for both the difference equations in the SM method, and the factorization problem in the WH technique.

3.3. WH functional equations in the angular region filled by a homogeneous isotropic medium in Electromagnetics

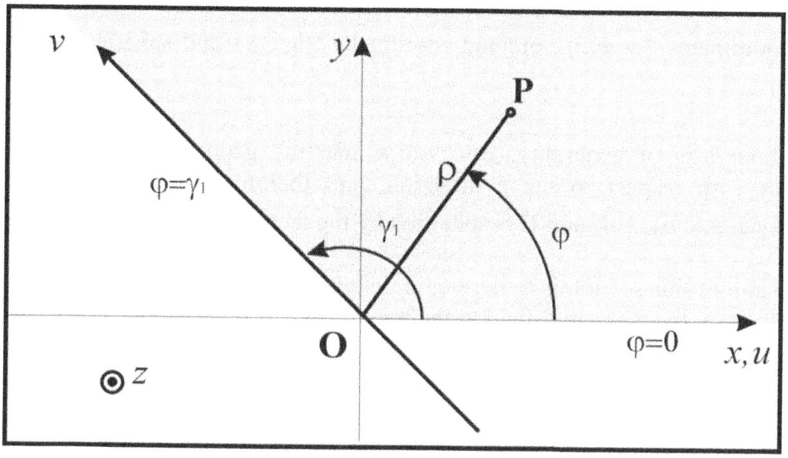

Figure 3.3.1. *Angular region* $0 \leq \varphi \leq \gamma_1 \leq \pi$ *with three coordinate systems: x,y,z Cartesian coordinates,* ρ,φ,z *cylindrical coordinates and u,v,z oblique Cartesian coordinates*

In the following, we only consider time harmonic electromagnetic fields with a time dependence specified by the factor $e^{j\omega t}$ which is omitted.

The deduction of the generalized Wiener-Hof equations (GWHEs) for an arbitrary angular region (Figure 3.3.1) with the aperture angle γ_1, $(0 \leq \varphi \leq \gamma_1)$ filled by a homogeneous isotropic medium involves two functional equations that relate Laplace transforms of fields tangential to the boundaries (Daniele, 2001; 2002; 2003b; 2004; Daniele and Lombardi, 2005; Daniele and Lombardi, 2016; Daniele *et al.* 2017a,b,c). Extension can be used for the angular region filled by an arbitrary homogeneous linear medium (Daniele, 2003b; 2004; Daniele and Lombardi, 2005; Daniele and Lombardi, 2016; Daniele *et al.* 2017a,b,c).

In this section, we demonstrate, step by step, the functional equations for a homogeneous isotropic medium. In this case, we have

$$\mathbf{D} = \varepsilon \mathbf{E}, \quad \mathbf{B} = \mu \mathbf{H} \tag{3.3.1}$$

where **D, B** are the induction fields, **E, H** are the electromagnetic fields, and ε and μ are respectively the scalar electric permittivity and magnetic permeability.

In order to describe the problem, we will use Cartesian coordinates (x, y, z), polar coordinates (ρ, φ, z), oblique coordinates (u, v, z) and spherical coordinates (r, φ, ϑ).

Without loss of generality, we assume that the geometry of the problem is invariant with respect to the z direction, and therefore the dependence of the electromagnetic fields **E** and **H** is specified by the factor $e^{-j\alpha_o z}$ which is omitted.

The aim of this section is to derive, in detail, a functional equation that relates in algebraic form the tangential field at the boundary $\varphi = 0$ to the one at the boundary $\varphi = \gamma_1$, as done in (3.3.46), from the wave equation in E_z. A similar functional equation can also be derived for H_z.

We introduce the following Laplace transforms:

$$V_{z+}(\eta,\varphi) = \int_0^\infty E_z(\rho,\varphi)e^{j\eta\rho}d\rho, \; I_{z+}(\eta,\varphi) = \int_0^\infty H_z(\rho,\varphi)e^{j\eta\rho}d\rho \tag{3.3.2}$$

$$V_{\rho+}(\eta,\varphi) = \int_0^\infty E_\rho(\rho,\varphi)e^{j\eta\rho}d\rho, \; I_{\rho+}(\eta,\varphi) = \int_0^\infty H_\rho(\rho,\varphi)e^{j\eta\rho}d\rho \tag{3.3.3}$$

where the subscript + indicates plus functions, i.e. functions having regular half-planes of convergence that are upper half-planes in the η-plane.

Note that although E_ρ and H_ρ may be singular for vanishing values of ρ, the existence of the Laplace transforms in physical problems is confirmed.

To avoid singular points on the real axis of the η-plane, we also assume small vanishing losses in the presence of non-dissipative media. Consequently, the propagation constants always have a negative imaginary part.

In the angular region, the field E_z must satisfy the wave equation (where we have omitted $e^{-j\alpha_o z}$):

$$\frac{\partial^2}{\partial x^2}E_z + \frac{\partial^2}{\partial y^2}E_z + \tau_o^2 E_z = 0 \tag{3.3.4}$$

with the wave number $k = \omega\sqrt{\varepsilon\mu}$, $\mathrm{Im}[k] \le 0$ and $\tau_o = \sqrt{k^2 - \alpha_o^2}$, $\mathrm{Im}[\tau_o] \le 0$. Note that, in the wedge scattering problem, α_o and τ_o are

$$\alpha_o = k\cos\beta \text{ and } \tau_o = k\sin\beta \tag{3.3.5}$$

which are related to the skewness of propagation, where $\vartheta = \beta$ is the zenithal angle in spherical coordinates.

Let us introduce the oblique Cartesian coordinates u and v to study this region $(\gamma = \gamma_1)$ (see Figure 3.3.1):

$$u = x - y\cot\gamma, \quad v = \frac{y}{\sin\gamma} \quad \text{or} \quad x = u + v\cos\gamma, \quad y = v\sin\gamma \tag{3.3.6}$$

with differentials

$$\frac{\partial}{\partial x} = \frac{\partial u}{\partial x}\frac{\partial}{\partial u} + \frac{\partial v}{\partial x}\frac{\partial}{\partial v} = \frac{\partial}{\partial u}, \qquad \frac{\partial}{\partial y} = \frac{\partial u}{\partial y}\frac{\partial}{\partial u} + \frac{\partial v}{\partial y}\frac{\partial}{\partial v} = -\cot\gamma\frac{\partial}{\partial u} + \frac{1}{\sin\gamma}\frac{\partial}{\partial v} \tag{3.3.7}$$

$$\frac{\partial}{\partial u} = \frac{\partial x}{\partial u}\frac{\partial}{\partial x} + \frac{\partial y}{\partial u}\frac{\partial}{\partial y} = \frac{\partial}{\partial x}, \qquad \frac{\partial}{\partial v} = \frac{\partial x}{\partial v}\frac{\partial}{\partial x} + \frac{\partial y}{\partial v}\frac{\partial}{\partial y} = \cos\gamma\frac{\partial}{\partial x} + \sin\gamma\frac{\partial}{\partial y} \tag{3.3.8}$$

thus

$$\frac{\partial^2}{\partial x^2} = \frac{\partial^2}{\partial u^2} \tag{3.3.9}$$

$$\begin{aligned}\frac{\partial^2}{\partial y^2} &= \frac{\partial}{\partial y}\left[-\cot\gamma\frac{\partial}{\partial u} + \frac{1}{\sin\gamma}\frac{\partial}{\partial v}\right] = \frac{\partial}{\partial y}\frac{\partial u}{\partial u}\left[-\cot\gamma\frac{\partial}{\partial u} + \frac{1}{\sin\gamma}\frac{\partial}{\partial v}\right] \\ &+ \frac{\partial}{\partial y}\frac{\partial v}{\partial v}\left[-\cot\gamma\frac{\partial}{\partial u} + \frac{1}{\sin\gamma}\frac{\partial}{\partial v}\right] = \frac{\partial}{\partial u}(-\cot\gamma)\left[-\cot\gamma\frac{\partial}{\partial u} + \frac{1}{\sin\gamma}\frac{\partial}{\partial v}\right] \\ &+ \frac{\partial}{\partial v}\left(\frac{1}{\sin\gamma}\right)\left[-\cot\gamma\frac{\partial}{\partial u} + \frac{1}{\sin\gamma}\frac{\partial}{\partial v}\right] = (\cot\gamma)^2\frac{\partial^2}{\partial u^2} + \frac{1}{(\sin\gamma)^2}\frac{\partial^2}{\partial v^2} - 2\frac{\cos\gamma}{(\sin\gamma)^2}\frac{\partial^2}{\partial u\partial v}\end{aligned} \tag{3.3.10}$$

With these coordinates, (3.3.4) assumes the form:

$$\frac{\partial^2 E_z}{\partial u^2}+\frac{\partial^2 E_z}{\partial v^2}-2\cos\gamma\frac{\partial^2 E_z}{\partial u\partial v}+\tau_0{}^2\sin^2\gamma\, E_z=0 \qquad (3.3.11)$$

Equation (3.3.11) is solved in the Laplace domain by introducing the Laplace transform:

$$\tilde{E}_z(\eta,v)=\mathrm{L}\left[E_z(u,v)\right]=\int_0^\infty E_z(u,v)e^{j\eta u}du \qquad (3.3.12)$$

It follows from (3.3.11) that:

$$\frac{d^2\,\tilde{E}_z}{dv^2}+2j\eta\cos\gamma\frac{d\,\tilde{E}_z}{dv}+\left(\tau_0^2\sin^2\gamma-\eta^2\right)\tilde{E}_z=f_\eta(v) \qquad (3.3.13)$$

where:

$$f_\eta(v)=-2\cos\gamma\frac{dE_z(0_+,v)}{dv}-j\eta E_z(0_+,v)+\left.\frac{\partial E_z(u,v)}{\partial u}\right|_{u=0_+} \qquad (3.3.14)$$

By applying the characteristic Green function procedure (Felsen and Marcuvitz 1973, p. 274; Friedman 1956, p.169), we obtain the solution of (3.3.13):

$$\tilde{E}_z(\eta,v)=A(\eta)e^{-j\bar{m}(\eta)v}-\int_0^\infty\frac{e^{-j\eta(v-v1)\cos\gamma}e^{-j\xi|v-v1|\sin\gamma}}{2j\sin\gamma\xi}f_\eta(v_1)\,dv_1 \qquad (3.3.15)$$

where $\xi=\xi(\eta)=\sqrt{\tau_o^2-\eta^2}$ with the branch $\xi(0)=\tau_o$ for $\eta=0$. A(η) is an unknown function and $\vec{m}(\eta)$ and $\bar{m}(\eta)$ are the two solutions of the characteristic equation related to (3.3.13):

$$-m^2+2\eta\cos\gamma\, m+\tau_o^2\sin^2\gamma-\eta^2=0 \qquad (3.3.16)$$

where $\frac{d}{dv}\Rightarrow -jm$. The solutions of (3.3.16) are

$$\vec{m}(\eta)=\cos\gamma\,\eta+\sin\gamma\,\xi\,,\qquad \bar{m}(\eta)=\cos\gamma\,\eta-\sin\gamma\,\xi \qquad (3.3.17)$$

and therefore the two solutions of the homogeneous equation associated with (3.3.13) are $\vec{\varphi}(v)=\exp(-j\vec{m}(\eta)v)$ and $\bar{\varphi}(v)=\exp(-j\bar{m}(\eta)v)$ and Green's function is

$$g(v,v_1)=\frac{\breve{\varphi}(v_<)\vec{\varphi}(v_>)}{W_r[\breve{\varphi}(v),\vec{\varphi}(v)]}=\frac{e^{-j\eta(v-v_1)\cos\gamma}e^{-j\xi|v-v_1|\sin\gamma}}{2j\sin\gamma\xi} \quad (3.3.18)$$

where W_r is the Wronskian and $v_>$ ($v_<$) is the greater (smaller) between v and v_1 that is the integration variable.

It yields (3.3.15), where $\breve{\varphi}(v)=\exp(-j\breve{m}(\eta)v)$ is not present out of integration due to the radiation condition at infinity. On the semi-axis v=0, (3.3.15) provides the equation:

$$V_{z+}(\eta,\varphi=0)=\tilde{E}_z(\eta,v=0)=A(\eta)-\int_0^\infty \frac{e^{-jm(\eta)v_1}}{2j\sin\gamma\xi}f_\eta(v_1)\,dv_1 \quad (3.3.19)$$

where:

$$m(\eta)=-\eta\cos\gamma+\sqrt{\tau_o^2-\eta^2}\sin\gamma=-\breve{m}(\eta) \quad (3.3.20)$$

The representation of the magnetic field H_x in the Laplace domain (3.3.28) can be obtained from Maxwell equations resulting from $H_x=\frac{1}{-j\omega\mu}\left(\frac{\partial E_z}{\partial y}-\frac{\partial E_y}{\partial z}\right)$ and using $E_y=\frac{1}{j\omega\varepsilon}\left(\frac{\partial H_x}{\partial z}-\frac{\partial H_z}{\partial x}\right)=\frac{1}{j\omega\varepsilon}\left(-j\alpha_o H_x-\frac{\partial H_z}{\partial x}\right)$, thus $\frac{\partial E_y}{\partial z}=$ $-j\alpha_o\frac{1}{j\omega\varepsilon}\left(-j\alpha_o H_x-\frac{\partial H_z}{\partial x}\right)$, we have

$$H_x=\frac{1}{-j\omega\mu}\left(\frac{\partial E_z}{\partial y}\right)+\frac{1}{-j\omega\mu}\left(j\alpha_o\frac{1}{j\omega\varepsilon}\left(-j\alpha_o H_x-\frac{\partial H_z}{\partial x}\right)\right)=\frac{1}{-j\omega\mu}\left(\frac{\partial E_z}{\partial y}\right)$$
$$+\frac{\alpha_o^2}{k^2}H_x-\frac{j\alpha_o}{k^2}\frac{\partial H_z}{\partial x} \quad (3.3.21)$$

After some mathematical manipulations, we have $\frac{\tau_o^2}{k^2}H_x=$ $\frac{1}{-j\omega\mu}\left(\frac{\partial E_z}{\partial y}\right)-\frac{j\alpha_o}{k^2}\frac{\partial H_z}{\partial x}$, which yields

$$H_x=\frac{k^2}{-j\omega\mu\tau_o^2}\left(\frac{\partial E_z}{\partial y}\right)-\frac{j\alpha_o}{\tau_o^2}\frac{\partial H_z}{\partial x}=\frac{j\omega\varepsilon}{\tau_o^2}\left(\frac{\partial E_z}{\partial y}\right)-\frac{j\alpha_o}{\tau_o^2}\frac{\partial H_z}{\partial x} \quad (3.3.22)$$

By applying (3.3.7) $\frac{\partial}{\partial y}=-\cot\gamma\frac{\partial}{\partial u}+\frac{1}{\sin\gamma}\frac{\partial}{\partial v}$ to (3.3.22), we have

$$H_x=-\frac{j\omega\varepsilon}{\tau_o^2}\left(-\frac{\partial E_z}{\partial v}\frac{1}{\sin\gamma}+\frac{\partial E_z}{\partial u}\cot\gamma\right)-\frac{j\alpha_o}{\tau_o^2}\frac{\partial H_z}{\partial u} \tag{3.3.23}$$

Using the Laplace transform

$$\tilde{H}_x(\eta,v)=\mathrm{L}\left[H_x(u,v)\right]=\int_0^\infty H_x(u,v)e^{j\eta u}du \tag{3.3.24}$$

in (3.3.23), it yields at $v=0$

$$I_{\rho+}(\eta,0)=\tilde{H}_x(\eta,0)=\frac{j\omega\varepsilon}{\tau_o^2}\frac{1}{\sin\gamma}\left(\left.\frac{\partial\tilde{E}_z(\eta,v)}{\partial v}\right|_{v=0}-(-j\eta\cos\gamma\tilde{E}_z(\eta,v=0)-\cos\gamma E_z(0,v=0))\right)$$
$$-\frac{j\alpha_o}{\tau_o^2}(-j\eta\tilde{H}_z(\eta,v=0)-H_z(0,v=0)) \tag{3.3.25}$$

By taking into account (3.3.15), we obtain

$$\frac{\partial\tilde{E}_z(\eta,v)}{\partial v}$$
$$=-j\vec{m}A(\eta)e^{-j\vec{m}(\eta)v}-\int_0^\infty\frac{e^{-j\eta(v-v_1)\cos\gamma}e^{-j\xi|v-v_1|\sin\gamma}}{2j\sin\gamma\xi}f_\eta(v_1)[-j\eta\cos\gamma-j\xi\sin\gamma sign(v-v_1)]dv_1 \tag{3.3.26}$$

and thus

$$\left.\frac{\partial\tilde{E}_z(\eta,v)}{\partial v}\right|_{v=0}=-j\vec{m}A(\eta)-jm\int_0^\infty\frac{e^{-j\eta v1\cos\gamma}e^{-j\xi v1\sin\gamma}}{2j\sin\gamma\xi}f_\eta(v_1)dv_1 \tag{3.3.27}$$

where m is defined in (3.3.20). Substituting (3.3.27) into (3.3.25), we obtain

$$I_{\rho+}(\eta,0)=\tilde{H}_x(\eta,0)$$
$$=\frac{j\omega\varepsilon}{\tau_o^2}\frac{1}{\sin\gamma}\left(-j\vec{m}A(\eta)-\frac{jm}{2j\sin\gamma\xi}\int_0^\infty e^{-jmv}f_\eta(v)dv+j\eta\cos\gamma\tilde{E}_z(\eta,0)+\cos\gamma E_z(0,0)\right)$$
$$-\frac{j\alpha_o}{\tau_o^2}(-j\eta I_{z+}(\eta)-H_z(0,0)) \tag{3.3.28}$$

Substituting (3.3.19) into (3.3.28), we eliminate the unknown function $A(\eta)$:

$$I_{\rho+}(\eta,0) = \tilde{H}_x(\eta,0) = \frac{j\omega\varepsilon}{\tau_o^2}\frac{1}{\sin\gamma}$$

$$\left(-j\vec{m}V_{z+}(\eta) - j\vec{m}\int_0^\infty \frac{e^{-jm(\eta)v_1}}{2j\sin\gamma\xi} f_\eta(v_1)\,dv_1 - \frac{jm}{2j\sin\gamma\xi}\int_0^\infty e^{-jmv} f_\eta(v)\,dv + j\eta\cos\gamma V_{z+}(\eta)\right.$$

$$\left. + \cos\gamma E_z(0,0)\right) - \frac{j\alpha_o}{\tau_o^2}(-j\eta I_{z+}(\eta) - H_z(0,0)) \tag{3.3.29}$$

yielding

$$I_{\rho+}(\eta,0) = \frac{j\omega\varepsilon}{\tau_o^2}\frac{1}{\sin\gamma}\left(-j\xi\sin\gamma V_{z+}(\eta) - \int_0^\infty e^{-jmv} f_\eta(v)\,dv + \cos\gamma E_z(0,0)\right) - \frac{j\alpha_o}{\tau_o^2}(-j\eta I_{z+}(\eta) - H_z(0,0)) \tag{3.3.30}$$

since $-j\vec{m} - jm = -2j\sin\gamma\xi$ and $-j\vec{m} + j\eta\cos\gamma = -j\xi\sin\gamma$. We obtain

$$I_{\rho+}(\eta,0) = \frac{\omega\varepsilon}{\tau_o^2}\xi V_{z+}(\eta)$$

$$-\frac{j\omega\varepsilon}{\tau_o^2}\frac{1}{\sin\gamma}\int_0^\infty e^{-jmv} f_\eta(v)\,dv + \frac{j\omega\varepsilon}{\tau_o^2}\frac{1}{\sin\gamma}\cos\gamma E_z(0,0) + \frac{j\alpha_o}{\tau_o^2} j\eta I_{z+}(\eta) + \frac{j\alpha_o}{\tau_o^2} H_z(0,0) \tag{3.3.31}$$

where we need to find an explicit form of the Laplace transform of $f_\eta(v)$ (3.3.14) in v. In particular, we focus on the representation of $\left.\frac{\partial E_z(u,v)}{\partial u}\right|_{u=0_+}$.

After some algebraic manipulations of the Maxwell equations written in the coordinates u and v, it yields

$$H_v = \mathbf{H}\bullet\hat{v} = H_x\hat{x}\bullet\hat{v} + H_y\hat{y}\bullet\hat{v} = \cos\gamma H_x + \sin\gamma H_y$$

$$= \cos\gamma\left(\frac{1}{-j\omega\mu}\left(\frac{\partial E_z}{\partial y} - \frac{\partial E_y}{\partial z}\right)\right) + \sin\gamma\left(\frac{1}{-j\omega\mu}\left(\frac{\partial E_x}{\partial z} - \frac{\partial E_z}{\partial x}\right)\right) \tag{3.3.32}$$

$$
\begin{aligned}
H_v &= \frac{1}{-j\omega\mu}\left[\cos\gamma\frac{\partial E_z}{\partial y} - \sin\gamma\frac{\partial E_z}{\partial x} - \cos\gamma\frac{\partial E_y}{\partial z} + \sin\gamma\frac{\partial E_x}{\partial z}\right] \\
&= \frac{1}{-j\omega\mu}\left[\cos\gamma\left(-\cot\gamma\frac{\partial E_z}{\partial u} + \frac{1}{\sin\gamma}\frac{\partial E_z}{\partial v}\right) - \sin\gamma\frac{\partial E_z}{\partial u} - \cos\gamma\frac{\partial E_y}{\partial z} + \sin\gamma\frac{\partial E_x}{\partial z}\right]
\end{aligned} \tag{3.3.33}
$$

$$
H_v = \frac{1}{-j\omega\mu}\left[-\frac{1}{\sin\gamma}\frac{\partial E_z}{\partial u} + \cot\gamma\frac{\partial E_z}{\partial v} + j\alpha_o\cos\gamma E_y - j\alpha_o\sin\gamma E_x\right] \tag{3.3.34}
$$

$$
\begin{aligned}
H_v = \frac{1}{-j\omega\mu}&\left[-\frac{1}{\sin\gamma}\frac{\partial E_z}{\partial u} + \cot\gamma\frac{\partial E_z}{\partial v} + j\alpha_o\cos\gamma\left(\frac{1}{j\omega\varepsilon}\left(\frac{\partial H_x}{\partial z} - \frac{\partial H_z}{\partial x}\right)\right)\right. \\
&\left. -j\alpha_o\sin\gamma\frac{1}{j\omega\varepsilon}\left(\frac{\partial H_z}{\partial y} - \frac{\partial H_y}{\partial z}\right)\right]
\end{aligned} \tag{3.3.35}
$$

$$
\begin{aligned}
H_v = \frac{1}{-j\omega\mu}&\left[-\frac{1}{\sin\gamma}\frac{\partial E_z}{\partial u} + \cot\gamma\frac{\partial E_z}{\partial v} + \frac{j\alpha_o\cos\gamma}{j\omega\varepsilon}(-j\alpha_o H_x)\right. \\
&\left. + \frac{j\alpha_o\cos\gamma}{j\omega\varepsilon}\left(-\frac{\partial H_z}{\partial x}\right) + \frac{j\alpha_o\sin\gamma}{j\omega\varepsilon}(-j\alpha_o H_y) - \frac{j\alpha_o\sin\gamma}{j\omega\varepsilon}\left(\frac{\partial H_z}{\partial y}\right)\right]
\end{aligned} \tag{3.3.36}
$$

$$
\begin{aligned}
H_v = \frac{1}{-j\omega\mu}&\left[-\frac{1}{\sin\gamma}\frac{\partial E_z}{\partial u} + \cot\gamma\frac{\partial E_z}{\partial v} - \frac{j\alpha_o^2\cos\gamma}{\omega\varepsilon}H_x - \frac{\alpha_o\cos\gamma}{\omega\varepsilon}\left(\frac{\partial H_z}{\partial u}\right)\right. \\
&\left. - \frac{j\alpha_o^2\sin\gamma}{\omega\varepsilon}H_y - \frac{\alpha_o\sin\gamma}{\omega\varepsilon}\left(-\cot\gamma\frac{\partial H_z}{\partial u} + \frac{1}{\sin\gamma}\frac{\partial H_z}{\partial v}\right)\right]
\end{aligned} \tag{3.3.37}
$$

and thus

$$
H_v = \frac{1}{-j\omega\mu}\left[-\frac{1}{\sin\gamma}\frac{\partial E_z}{\partial u} + \cot\gamma\frac{\partial E_z}{\partial v} - \frac{j\alpha_o^2\cos\gamma}{\omega\varepsilon}H_x - \frac{j\alpha_o^2\sin\gamma}{\omega\varepsilon}H_y - \frac{\alpha_o}{\omega\varepsilon}\frac{\partial H_z}{\partial v}\right] \tag{3.3.38}
$$

Since, at u=0 (along the v axis), we have $H_x = H_v\cos\gamma, H_y = H_v\sin\gamma$, (3.3.38) becomes

$$
H_v = \frac{1}{-j\omega\mu}\left[-\frac{1}{\sin\gamma}\frac{\partial E_z}{\partial u} + \cot\gamma\frac{\partial E_z}{\partial v} - \frac{j\alpha_o^2}{\omega\varepsilon}H_v - \frac{\alpha_o}{\omega\varepsilon}\frac{\partial H_z}{\partial v}\right] \tag{3.3.39}
$$

from which we obtain

$$\left.\frac{\partial E_z}{\partial u}\right|_{0_+} = \frac{\sin\gamma}{\omega\varepsilon}\left(j\tau_o^2 H_v + \omega\varepsilon \cot\gamma \frac{\partial E_z}{\partial v} - \alpha_o \frac{\partial H_z}{\partial v} \right) \tag{3.3.40}$$

As stated previously, we need to perform the Laplace transform of $f_\eta(v)$ (3.3.14) in v; therefore, from the definition of $f_\eta(v)$ and using (3.3.40), we have

$$f_\eta(v) = \cos\gamma \frac{\partial E_z(0_+, v)}{\partial v} - j\eta E_z(0_+, v) - \alpha_o \frac{\sin\gamma}{\omega\varepsilon}\frac{\partial H_z}{\partial v} + j\tau_o^2 \frac{\sin\gamma}{\omega\varepsilon} H_v \tag{3.3.41}$$

Since

$$\begin{aligned} &\int_0^\infty e^{j\eta u} f'(u)\,du = e^{j\eta u} f(u)\Big|_0^\infty - (j\eta)\int_0^\infty e^{j\eta u} f(u)\,du = -j\eta F(\eta) - f(0) \\ &\int_0^\infty e^{-jmv} f'(v)\,dv = e^{-jmv} f(v)\Big|_0^\infty - (-jm)\int_0^\infty e^{-jmv} f(v)\,dv = jmF(-m) - f(0) \end{aligned} \tag{3.3.42}$$

we have

$$\begin{aligned} \int_0^\infty e^{-jmv} f_\eta(v)\,dv &= -\cos\gamma\left(jmV_{z+}(-m,\gamma) - E_z(0,0)\right) - j\eta V_{z+}(-m,\gamma) \\ &- \frac{\alpha_o \sin\gamma}{\omega\varepsilon}\left(jmI_{z+}(-m,\gamma) - H_z(0,0)\right) + \frac{j\tau_o^2 \sin\gamma}{\omega\varepsilon} I_{\rho+}(-m,\gamma) \end{aligned} \tag{3.3.43}$$

By manipulating (3.3.31)

$$\begin{aligned} &\xi V_{z+}(\eta) - \frac{\tau_o^2}{\omega\varepsilon} I_{\rho+}(\eta,0) - \frac{\alpha_o \eta}{\omega\varepsilon} I_{z+}(\eta) = \frac{j}{\sin\gamma}\int_0^\infty e^{-jmv} f_\eta(v)\,dv \\ &- j\frac{\cos\gamma}{\sin\gamma} E_z(0,0) - \frac{j\alpha_o}{\omega\varepsilon} H_z(0,0) \end{aligned} \tag{3.3.44}$$

and substituting (3.3.43), we have

$$\begin{aligned} &\xi V_{z+}(\eta) - \frac{\tau_o^2}{\omega\varepsilon} I_{\rho+}(\eta,0) - \frac{\alpha_o \eta}{\omega\varepsilon} I_{z+}(\eta) = \frac{j}{\sin\gamma}\Big[-\cos\gamma\, jmV_{z+}(-m,\gamma) - j\eta V_{z+} \\ &(-m,\gamma) - \frac{\alpha_o \sin\gamma}{\omega\varepsilon} jmI_{z+}(-m,\gamma) + \frac{j\tau_o^2 \sin\gamma}{\omega\varepsilon} I_{\rho+}(-m,\gamma) + \cos\gamma E_z(0,0) + \frac{\alpha_o \sin\gamma}{\omega\varepsilon} H_z(0,0) \Big] \\ &- j\frac{\cos\gamma}{\sin\gamma} E_z(0,0) - \frac{j\alpha_o}{\omega\varepsilon} H_z(0,0) \end{aligned} \tag{3.3.45}$$

and finally

$$\xi V_{z+}(\eta,0)-\frac{\tau_o^2}{\omega\varepsilon}I_{\rho+}(\eta,0)-\frac{\alpha_o\eta}{\omega\varepsilon}I_{z+}(\eta)$$
$$=-nV_{z+}(-m,\gamma)-\frac{\tau_o^{\,2}}{\omega\varepsilon}I_{\rho+}(-m,\gamma)+\left(\frac{\alpha_o m}{\omega\varepsilon}\right)I_{z+}(-m,\gamma)$$

(3.3.46)

where:

$$n=n(\eta)=-\xi\cos\gamma-\eta\sin\gamma=\sqrt{\tau_o^2-m^2}$$
$$m=m(\eta)=-\eta\cos\gamma+\xi\sin\gamma \qquad (3.3.47)$$

and $\dfrac{(\cos\gamma m+\eta)}{\sin\gamma}=\dfrac{(\cos\gamma(-\eta\cos\gamma+\xi\sin\gamma)+\eta)}{\sin\gamma}=\xi\cos\gamma+\eta\sin\gamma=-n$

As stated at the beginning of this section, (3.3.46) is a functional equation that relates field components at $\varphi=0$ to the ones at $\varphi=\gamma$ resulting from the wave equation in E_z (3.3.4).

Following the same procedure, a functional equation is also derived from the wave equation for H_z:

$$\frac{\partial^2}{\partial x^2}H_z+\frac{\partial^2}{\partial y^2}H_z+\tau_o^2 H_z=0 \qquad (3.3.48)$$

It yields

$$\xi I_{z+}(\eta,0)+\frac{\tau_o^2}{\omega\mu}V_{\rho+}(\eta,0)+\frac{\alpha_o\eta}{\omega\mu}V_{z+}(\eta,0)=-nI_{z+}(-m,\gamma)+\frac{\tau_o^2}{\omega\mu}V_{\rho+}(-m,\gamma)-\frac{\alpha m}{\omega\mu}V_{z+}(-m,\gamma)$$

(3.3.49)

Alternatively, we can obtain (3.3.49) by the duality principle; in fact, we observe substituting $\mathrm{E}\rightarrow\mathrm{H}$, $\mu\rightarrow-\varepsilon$ and the first Maxwell equation becomes the second one and vice versa.

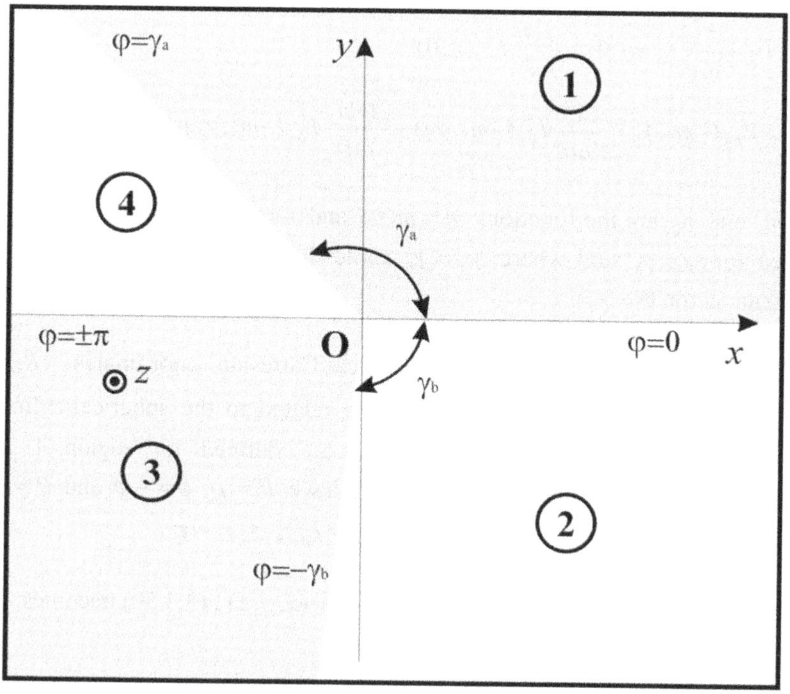

Figure 3.3.2. *Space divided into four angular regions delimited by* $\varphi = \gamma_a, -\gamma_b, \pm\pi$

In order to study complex problems that involve angular regions, we derive functional equations for the tangential components of electromagnetic fields in regions 2, 3 and 4 from those of region 1 $0<\varphi<\gamma_a=\gamma_1$, i.e. (3.3.46) and (3.3.49) with $\gamma_1=\gamma$, by using symmetries, duality and extrapolation.

Let us modify the above procedure to analyze region 2 of Figure 3.3.2 $-\gamma_b<\varphi<0$ consisting of the same material of region 1.

In particular, we use the new Cartesian axes defined by $X=x$, $Y=-y$, $Z=-z$ and a related set of cylindrical and spherical coordinates respectively (P,ϕ,Z) and (R,ϕ,θ).

With (X,Y,Z) for region 2 with the aperture angle $\gamma_b>0$, we obtain a functional equation that is slightly modified with respect to (3.3.46) from the wave equation in E_z:

$$\xi V_{Z+}(\eta,0)-\frac{\tau_o^2}{\omega\varepsilon}I_{P+}(\eta,0)-\frac{A_o\eta}{\omega\varepsilon}I_{Z+}(\eta,0)$$
$$=-n_2\ V_{Z+}(-m_2,\gamma_B)-\frac{\tau_o^2}{\omega\varepsilon}I_{P+}(-m_2,\gamma_B)+\frac{A_o\,m_2}{\omega\varepsilon}I_{Z+}(-m_2,\gamma_B) \quad (3.3.50)$$

where m_2 and n_2 are the functions $m=m(\eta)$ and $n=n(\eta)$ defined in (3.3.47) and evaluated for $\gamma=\gamma_b$ and where $\gamma_B=\gamma_b$ (note that the direction of oblique field components is for $\phi=\gamma_B$).

We observe that the spherical/cylindrical/Cartesian coordinates $(R,\phi,\theta)/(P,\phi,Z)/(X,Y,Z)$ defined for region 2 are related to the spherical/cylindrical/Cartesian coordinates $(r,\varphi,\vartheta)/(\rho,\varphi,z)/(x,y,z)$ defined for region 1 at the beginning of this section, and, in particular, we have $P=\rho$, $\phi=-\varphi$ and $\theta=\pi-\vartheta$, $Z=-z$ therefore $V_{Z+}=-V_{z+}$, $I_{Z+}=-I_{z+}$, $I_{P+}=I_{\rho+}$, $A_o=-\alpha_o$.

Consequently, using the reference coordinates (x,y,z), (3.3.50) becomes:

$$\xi\ V_{z+}(\eta,0)+\frac{\tau_o^2}{\omega\varepsilon}I_{\rho+}(\eta,0)+\frac{\alpha_o\eta}{\omega\varepsilon}I_{z+}(\eta,0)$$
$$=-n_2\,V_{z+}(-m_2,-\gamma_b)+\frac{\tau_o^2}{\omega\varepsilon}I_{\rho+}(-m_2,-\gamma_b)-\frac{\alpha_o\,m_2}{\omega\varepsilon}I_{z+}(-m_2,-\gamma_b) \quad (3.3.51)$$

where the direction of oblique field components are now for $\varphi=-\gamma_b$.

Similarly, equation (3.3.49) for region 2 becomes:

$$\xi\ I_{z+}(\eta,0)-\frac{\tau_o^2}{\omega\mu}V_{\rho+}(\eta,0)-\frac{\alpha_o\eta}{\omega\mu}V_{z+}(\eta,0)$$
$$=-n\ I_{z+}(-m,-\gamma_b)-\frac{\tau_o^2}{\omega\mu}V_{\rho+}(-m,-\gamma_b)+\frac{\alpha_o\,m}{\omega\mu}V_{z+}(-m,-\gamma_b) \quad (3.3.52)$$

Note that the equations obtained in each region also hold whatever medium fills the other regions. Consequently, in order to obtain the functional equation for region 4, we can assume that region 1 is filled by the same medium of region 4.

By putting $\gamma_a=\pi$ and assuming the parameters of medium 4, (3.3.46) and (3.3.49) become:

$$\xi\, V_{z+}(\eta,0)-\frac{\tau_o^2}{\omega\varepsilon}I_{\rho+}(\eta,0)-\frac{\alpha_o\eta}{\omega\varepsilon}I_{z+}(\eta,0)=-\xi\, V_{z+}(-\eta,\pi)-\frac{\tau_o^2}{\omega\varepsilon}I_{\rho+}(-\eta,\pi)+\frac{\alpha_o\eta}{\omega\varepsilon}I_{z+}(-\eta,\pi) \tag{3.3.53}$$

$$\xi\, I_{z+}(\eta,0)+\frac{\tau_o^2}{\omega\mu}V_{\rho+}(\eta,0)+\frac{\alpha_o\eta}{\omega\mu}V_{z+}(\eta,0)=-\xi\, I_{z+}(-\eta,\pi)+\frac{\tau_o^2}{\omega\mu}V_{\rho+}(-\eta,\pi)-\frac{\alpha\eta}{\omega\mu}V_{z+}(-\eta,\pi) \tag{3.3.54}$$

Substituting in (3.3.53) and (3.3.54) equations (3.3.46) and (3.3.49) with the parameters of medium 4, we eliminate the first members in the equations (the field components for $\varphi=0$), yielding:

$$\begin{aligned}&-\xi\, V_{z+}(-\eta,\pi)-\frac{\tau_o^2}{\omega\varepsilon}I_{\rho+}(-\eta,\pi)+\frac{\alpha_o\eta}{\omega\varepsilon}I_{z+}(-\eta,\pi)\\&=-n\, V_{z+}(-m,\gamma_a)-\frac{\tau_o^2}{\omega\varepsilon}I_{\rho+}(-m,\gamma_a)+\frac{\alpha_o m}{\omega\varepsilon}I_{z+}(-m,\gamma_a)\end{aligned} \tag{3.3.55}$$

$$\begin{aligned}&-\xi\, I_{z+}(-\eta,\pi)+\frac{\tau_o^2}{\omega\mu}V_{\rho+}(-\eta,\pi)-\frac{\alpha_o\eta}{\omega\mu}V_{z+}(-\eta,\pi)\\&=-n\, I_{z+}(-m,\gamma_a)+\frac{\tau_o^2}{\omega\mu}V_{\rho+}(-m,\gamma_a)-\frac{\alpha m}{\omega\mu}V_{z+}(-m,\gamma_a)\end{aligned} \tag{3.3.56}$$

The same procedure holds for region 3 by using equations of region 2.

According to Figure 3.3.2, we have in the general case:

Region 1 $(0\le\varphi\le\gamma_a)$

$$\begin{aligned}&\xi_1 V_{z+}(\eta,0)-\frac{\tau_{o1}^2}{\omega\varepsilon_1}I_{\rho+}(\eta,0)-\frac{\alpha_{o1}\eta}{\omega\varepsilon_1}I_{z+}(\eta,0)\\&=-n_1 V_{z+}(-m_1,\gamma_a)-\frac{\tau_{o1}^2}{\omega\varepsilon_1}I_{\rho+}(-m_1,\gamma_a)+\frac{\alpha_{o1}m_1}{\omega\varepsilon_1}I_{z+}(-m_1,\gamma_a)\end{aligned} \tag{3.3.57}$$

$$\begin{aligned}&\xi_1 I_{z+}(\eta,0)+\frac{\tau_{o1}^2}{\omega\mu_1}V_{\rho+}(\eta,0)+\frac{\alpha_{o1}\eta}{\omega\mu_1}V_{z+}(\eta,0)\\&=-n_1 I_{z+}(-m_1,\gamma_a)+\frac{\tau_{o1}^2}{\omega\mu_1}V_{\rho+}(-m1,\gamma_a)-\frac{\alpha_{o1}m_1}{\omega\mu_1}V_{z+}(-m_1,\gamma_a)\end{aligned} \tag{3.3.58}$$

Region 2 $(-\gamma_b \le \varphi \le 0)$

$$\begin{aligned}&\xi_2 V_{z+}(\eta,0)+\frac{\tau_{o2}^2}{\omega\varepsilon_2}I_{\rho+}(\eta,0)+\frac{\alpha_{o2}\eta}{\omega\varepsilon_2}I_{z+}(\eta,0)\\&=-n_2 V_{z+}(-m_2,-\gamma_b)+\frac{\tau_{o2}^2}{\omega\varepsilon_2}I_{\rho+}(-m_2,-\gamma_b)-\frac{\alpha_{o2}m_2}{\omega\varepsilon_2}I_{z+}(-m_2,-\gamma_b)\end{aligned}\tag{3.3.59}$$

$$\begin{aligned}&\xi_2 I_{z+}(\eta,0)-\frac{\tau_{o2}^2}{\omega\mu_2}V_{\rho+}(\eta,0)-\frac{\alpha_{o2}\eta}{\omega\mu_2}V_{z+}(\eta,0)\\&=-n_2 I_{z+}(-m_2,-\gamma_b)-\frac{\tau_{o2}^2}{\omega\mu_2}V_{\rho+}(-m_2,-\gamma_b)+\frac{\alpha_{o2}m_2}{\omega\mu_2}V_{z+}(-m_2,-\gamma_b)\end{aligned}\tag{3.3.60}$$

Region 3 $(-\pi \le \varphi \le -\gamma_b)$

$$\begin{aligned}&-\xi_3 V_{z+}(-\eta,-\pi)+\frac{\tau_{03}^2}{\omega\varepsilon_3}I_{\rho+}(-\eta,-\pi)-\frac{\alpha_{03}\eta}{\omega\varepsilon_3}I_{z+}(-\eta,-\pi)\\&=-n_3 V_{z+}(-m_3,-\gamma_b)+\frac{\tau_{03}^2}{\omega\varepsilon_3}I_{\rho+}(-m_3,-\gamma_b)-\frac{\alpha_{o3}m_3}{\omega\varepsilon_3}I_{z+}(-m_3,-\gamma_b)\end{aligned}\tag{3.3.61}$$

$$\begin{aligned}&-\xi_3 I_{z+}(-\eta,-\pi)-\frac{\tau_{o3}^2}{\omega\mu_3}V_{\rho+}(-\eta,-\pi)+\frac{\alpha_{o3}\eta}{\omega\mu_3}V_{z+}(-\eta,-\pi)\\&=-n_3 I_{z+}(-m_3,-\gamma_b)-\frac{\tau_{o3}^2}{\omega\mu_3}V_{\rho+}(-m_3,-\gamma_b)+\frac{\alpha_{o3}m_3}{\omega\mu_3}V_{z+}(-m_3,-\gamma_b)\end{aligned}\tag{3.3.62}$$

Region 4 $(\gamma_a \le \varphi \le \pi)$

$$\begin{aligned}&\xi_4 V_{z+}(-\eta,\pi)+\frac{\tau_{o4}^2}{\omega\varepsilon_4}I_{z+}(-\eta,\pi)-\frac{\alpha_{o4}\eta}{\omega\varepsilon_4}I_{z+}(-\eta,\pi)\\&=n_4 V_{z+}(-m_4,\gamma_a)+\frac{\tau_{o4}^2}{\omega\varepsilon_4}I_{\rho+}(-m_4,\gamma_a)-\frac{\alpha_{o4}m_4}{\omega\varepsilon_4}I_{z+}(-m_4,\gamma_a)\end{aligned}\tag{3.3.63}$$

$$\begin{aligned}&-\xi_4 I_{z+}(-\eta,\pi)+\frac{\tau_{o4}^2}{\omega\mu_4}V_{\rho+}(-\eta,\pi)-\frac{\alpha_{o4}\eta}{\omega\mu_4}V_{z+}(-\eta,\pi)\\&=-n_4 I_{z+}(-m_4,\gamma_a)+\frac{\tau_{o4}^2}{\omega\mu_4}V_{\rho+}(-m_4,\gamma_a)-\frac{\alpha_{o4}m_4}{\omega\mu_4}V_{z+}(-m_4,\gamma_a)\end{aligned}\tag{3.3.64}$$

In (3.3.57)–(3.3.64), the parameters $\xi_i, \tau_i, \varepsilon_i, \mu_i, m_i, n_i \quad i = 1,2,3,4$ are relevant to the considered region respectively with $\gamma_i = \{\gamma_a, \gamma_b, \gamma_b, \gamma_a\}$ and

$$\begin{aligned}
&\xi_i = \sqrt{\tau_{oi}^2 - \eta^2}, \quad \tau_{oi} = \sqrt{k_i^2 - \alpha_{oi}^2}, \quad k_i = \omega\sqrt{\varepsilon_i \mu_i} \\
&m_i = -\eta \cos\gamma_i + \xi_i \sin\gamma_i, \quad n_i = -\xi_i \cos\gamma_i - \eta \sin\gamma_i = \sqrt{\tau_{oi}^2 - m_i^2}
\end{aligned} \tag{3.3.65}$$

As expected in section 1.4, it is worth observing that, in the case of scattering problems due to plane waves, the relevant WH functional equations are homogeneous.

In the context of angular regions, we can verify that the incident plane wave defined by

$$\begin{aligned}
&E_z^i = E_o e^{j\tau_o \rho \cos(\varphi - \varphi_o)} e^{-j\alpha_o z}, \quad H_z^i = H_o e^{j\tau_o \rho \cos(\varphi - \varphi_o)} e^{-j\alpha_o z} \\
&E_\rho^i = \frac{1}{j\tau_o^2}\left[\alpha_o \frac{\partial E_z^i}{\partial \rho} + \frac{kZ_o}{\rho}\frac{\partial H_z^i}{\partial \varphi}\right], \quad H_\rho^i = \frac{1}{j\tau_o^2}\left[\alpha_0 \frac{\partial H_z^i}{\partial \rho} - \frac{k}{Z_o \rho}\frac{\partial E_z^i}{\partial \varphi}\right]
\end{aligned} \tag{3.3.66}$$

where $\alpha_o = k\cos\beta$ and $\tau_o = k\sin\beta$ with the skewness zenithal angle $\vartheta = \beta$ and the azimuthal angle φ_o, satisfy the equations of region 1 (3.3.46) and (3.3.49).

Hence, equations (3.3.57)–(3.3.64), derived in the absence of sources of scattering fields, also hold in the presence of plane wave sources as homogeneous WH functional equations. Thus, (3.3.57)–(3.3.64) are valid while the spectral functions represent scattering fields, total fields or plane wave sources.

We note that (3.3.57)–(3.3.64) are functional equations in the framework of the WH technique in a generalized form because they contain plus and minus functions in different complex planes η and m_i, where m_i are related to η through (3.3.65). We denote (3.3.57)–(3.3.64) as the generalized Wiener-Hopf functional equations for angular regions. In the following section, we will show that for every angular region, a suitable map exists, which reduces the plus and minus functions defined in different complex planes to the plus and minus functions defined in the same complex plane.

3.4. Reduction of the generalized functional equations of an angular region to functional equations defined in the same complex plane

In this section, we limit the procedure to plus and minus functions introduced in angular region 1 (3.3.57) and (3.3.58). Similar considerations apply for the functions of the other regions (3.3.59)–(3.3.64). For the sake of simplicity, we will omit subscript 1 and denote the aperture angle γ_1 of angular region 1 by Φ.

Let us consider a generic plus function $X_+(\eta)$ and a generic minus function $Y_+(-m) = X_-(m)$ of the generalized functional equations of region 1 (3.3.57) and (3.3.58). We show that the mapping $\eta = \eta(\bar{\eta})$ defined by

$$\eta = \eta(\bar{\eta}) = -\tau_o \cos\left[\frac{\Phi}{\pi}\left[\arccos\left[-\frac{\bar{\eta}}{\tau_o}\right]\right]\right],$$

$$\bar{\eta} = \bar{\eta}(\eta) = -\tau_o \cos\left[\frac{\pi}{\Phi}\left[\arccos\left[-\frac{\eta}{\tau_o}\right]\right]\right] \tag{3.4.1}$$

produces the functions $X_+(\eta(\bar{\eta})) = \bar{X}_+(\bar{\eta})$ and $X_-(m(\bar{\eta})) = \bar{X}_-(\bar{\eta})$ that are respectively plus and minus functions of the new plane $\bar{\eta}$.

In order to simplify functional–theoretical manipulations, we introduce the angular complex variable w, defined in section 1.13:

$$\eta = -\tau_o \cos w \tag{3.4.2}$$

In the w plane, we have

$$m = -\eta\cos\Phi + \xi\sin\Phi = \tau_o\cos(w+\Phi) \tag{3.4.3}$$

$$\xi = -\tau_o \sin(w) \tag{3.4.4}$$

Let us also introduce the variable

$$\bar{w} = \frac{\pi}{\Phi} w \text{ or } w = \frac{\Phi}{\pi}\bar{w} \tag{3.4.5}$$

It follows that the fundamental mapping (3.4.1) is equivalent to:

$$\bar{\eta} = -\tau_o \cos\bar{w} \tag{3.4.6}$$

and

$$m(\bar{\eta}) = -\eta(\bar{\eta})\cos\Phi + \sqrt{\tau_o^2 - \eta^2(\bar{\eta})}\sin\Phi \tag{3.4.7}$$

$$m = m(\bar{\eta}) = \tau_o \cos\left(\frac{\Phi}{\pi}\arccos\left(-\frac{\bar{\eta}}{\tau_o}\right) + \Phi\right) \tag{3.4.8}$$

Henceforth, we use the following notations for a plus function $f_+(\eta)$:

$$f_+(\eta) = f_+(-\tau_o \cos w) = \hat{f}_+(w) = \bar{f}_+(\bar{\eta}) = \bar{f}_+(-\tau_o \cos \bar{w}) = \tilde{f}_+(\bar{w}) \tag{3.4.9}$$

In order to demonstrate that $X_+(\eta(\bar{\eta})) = \bar{X}_+(\bar{\eta})$ and $Y_+(-m) = X_-(m(\bar{\eta})) = \bar{X}_-(\bar{\eta})$ are plus and minus functions in $\bar{\eta}$, we use the properties described in section 1.14, which are valid in the w plane for plus and minus functions defined by η, and we extend them to the complex plane $\bar{w}$ (3.4.5) by resorting to the relation between η and $\bar{\eta}$ (3.4.1).

We recall that the properties of the functions $f_+(\eta), f_-(\eta)$ redefined in the w plane are as follows:

1) plus functions of η are even functions of w, i.e. $\hat{f}_+(-w) = \hat{f}_+(w)$;

2) $\hat{f}_+(w)$ are regular at w=0, i.e. $\eta = -\tau_o$;

3) $f_-(\eta) = \hat{f}_-(w) = g_+(-\eta) = \hat{g}_+(-w-\pi)$ and, for the even property of plus functions, $\hat{f}_-(w) = \hat{g}_+(-w-\pi) = \hat{g}_+(w+\pi)$, thus a minus function does not change value in the w plane if w is substituted by $-2\pi - w$ and $\hat{f}_-(w)$ are regular at $w=-\pi$.

We also note that the mapping (3.4.5) (which is linear) keeps the symmetry of functions in complex angular planes from w- to $\bar{w}$ -plane and vice versa.

It follows that $\bar{X}_+(\bar{\eta}) = X_+(\eta)$ is a plus function in both η and $\bar{\eta}$ since $\tilde{X}_+(\bar{w})$ is even in $\bar{w}$ and regular at $\bar{w} = 0$ (which corresponds to $w = 0$, $\eta = -\tau_o$ and $\bar{\eta} = -\tau_o$). In fact

$$\bar{X}_+(\bar{\eta}) = \tilde{X}_+(\bar{w}) = X_+(-\tau_o \cos\frac{\Phi}{\pi}\bar{w}) = \hat{X}_+(w) = \hat{X}_+(-w) = X_+(-\tau_o \cos\frac{\Phi}{\pi}(-\bar{w})) = \tilde{X}_+(-\bar{w}) \tag{3.4.10}$$

$$\bar{X}_+(-\tau_o) = \tilde{X}_+(0) = X_+(-\tau_o) = \hat{X}_+(0) \tag{3.4.11}$$

and $X_+(\eta)$ is regular in $\eta = -\tau_o$.

In order to demonstrate that $Y_+(-m) = X_-(m(\bar{\eta})) = \bar{X}_-(\bar{\eta})$ is a minus function in both m and $\bar{\eta}$, we use property 3 and the definition of m in the $\bar{w}$ -plane:

$$m(\bar{w}) = \tau_o \cos\left(\frac{\Phi}{\pi}\bar{w} + \Phi\right) = \tau_o \cos\left(\frac{\Phi}{\pi}(\bar{w} + \pi)\right) \tag{3.4.12}$$

Property 3 ensures that a function with $\hat{f}_-(w) = \hat{g}_+(-w-\pi) = \hat{g}_+(w+\pi)$ and regular in $w=-\pi$ is a minus function in η. Since

$$\begin{aligned}\bar{X}_-(\bar{\eta}) &= \tilde{X}_-(\bar{w}) = \hat{X}_-(w) = Y_+(-m(\bar{\eta})) \\ &= Y_+(-\tau_o \cos\frac{\Phi}{\pi}(\bar{w} + \pi)) = \tilde{Y}_+(\bar{w} + \pi) = \tilde{Y}_+(-\bar{w} - \pi) = \hat{Y}_+(w + \Phi)\end{aligned} \tag{3.4.13}$$

and for $\bar{w} = -\pi$ (which corresponds to $m = \tau_o$ and $\bar{\eta} = \tau_o$), we have

$$\bar{X}_-(\tau_o) = \tilde{X}_-(-\pi) = Y_+(-\tau_o) = X_-(\tau_o) = \tilde{Y}_+(0) \tag{3.4.14}$$

with $X_-(m)$ being regular in $m = \tau_o$, it yields that $\bar{X}_-(\bar{\eta}) = \tilde{X}_-(\bar{w}) = Y_+(-m(\bar{\eta}))$ is a minus function in $\bar{\eta}$.

Note that the singularities involved in the $\bar{\eta}$ -plane only derive from the ones present in the η -plane (here we anticipate that the mapping does not introduce further singularities in the proper sheet). They consist of a) a finite number of poles/branches due to the structure (structural surface, lateral, leaky waves), b) branch points $\bar{\eta} = \pm\tau_o$ and c) a limited number of poles that arise from the geometrical optical (GO) field.

The limited number of poles involved allows us to conclude that $\bar{X}_+(\bar{\eta})$ has a regular upper half-plane and $\bar{Y}_+(-\bar{\eta}) = \bar{X}_-(\bar{\eta})$ has a regular lower half-plane.

It is remarkable to observe that the above properties also extend to the functions $p(\eta)X_+(\eta)$ and $q(m)Y_+(-m)$, where $p(\eta)$ and $q(m)$ are respectively polynomials of η and m, i.e. $\bar{p}(\bar{\eta})\bar{X}_+(\bar{\eta})$ and $\bar{q}(\bar{\eta})\bar{X}_-(\bar{\eta})$ are plus and minus functions in $\bar{\eta}$ although $\bar{p}(\bar{\eta})$ and $\bar{q}(\bar{\eta})$ are no longer polynomials.

Let us focus on the properties of the mapping (3.4.1). The mapping introduces into $\bar{X}_+(\bar{\eta})$ and $\bar{X}_-(\bar{\eta})$ the multivalued function:

$$\arccos\left(-\frac{\bar{\eta}}{\tau_o}\right) = -j\log\frac{\bar{\eta}+j\sqrt{\tau_o^2-\bar{\eta}^2}}{\tau_o} = -j\log\frac{\bar{\eta}+j\kappa(\bar{\eta})}{\tau_o} \tag{3.4.15}$$

Due to (3.4.15), in the $\bar{\eta}$-plane, we have the presence of the branch points and related branch lines only generated by $\kappa(\bar{\eta}) = \sqrt{\tau_o^2-\bar{\eta}^2}$, $\kappa(0) = \tau_o$, $\text{Im}[\tau_o] < 0$. Therefore, we need to transpose the concepts of proper/improper sheets (described in section 1.13 for η and w) in the new complex planes $\bar{\eta}$ and $\bar{w}$ with $\bar{\eta} = -\tau_o \cos\bar{w}$. Note also that $\bar{\xi}(\bar{\eta}) = \xi(\eta(\bar{\eta})) \neq \kappa(\bar{\eta})$.

Equation (3.4.15) is also related to the inversion of the mapping $\bar{\eta} = -\tau_o \cos\bar{w}$ (3.4.6):

$$\bar{w}(\bar{\eta}) = \arccos\left(-\frac{\bar{\eta}}{\tau_o}\right) \tag{3.4.16}$$

which introduces infinite replicates of $\bar{w}$.

For example, to achieve a unique relation between a point $\bar{\eta}$ and its corresponding point $\bar{w}$ from (3.4.16), we need to uniquely define the properties of the log function, as in section 1.13: for each complex number z, we assume $-\pi < \arg[z] \leq +\pi$ such that $\log(z) = \log|z| + \arg(z)$.

According to (3.4.15) and (3.4.16), we need to define the value of $\kappa(\bar{\eta}) = \sqrt{\tau_o^2-\bar{\eta}^2}$. In this section (as in section 1.13 for η and w), we select, unless otherwise specified, the branch lines consisting of the arcs of hyperbolas, where $\text{Im}[\kappa(\bar{\eta})] = 0$, from the branch points $\bar{\eta} = \pm\tau_o$ (Figure 3.4.1) and we assume the proper sheet with $\text{Im}[\kappa(\bar{\eta})] < 0$ and $\kappa(0) = \tau_o$. Alternative possible options for the branch lines are the vertical lines or $\text{Re}[\kappa(\bar{\eta})] = 0$ that define different proper/improper regions.

With the selected assumptions on the branch of $\kappa(\bar{\eta}) = \sqrt{\tau_o^2-\bar{\eta}^2}$ in the proper sheet and on the principal value of $\log(z)$, we have the general relationship:

$$\bar{w}(\bar{\eta}) = \begin{cases} -j\log\left(\frac{\bar{\eta}+j\kappa(\bar{\eta})}{\tau_o}\right) - \pi, & \arg\frac{\bar{\eta}+j\kappa(\bar{\eta})}{\tau_o} > -\frac{\pi}{2} \\ -j\log\left(\frac{\bar{\eta}+j\kappa(\bar{\eta})}{\tau_o}\right) + \pi, & \arg\frac{\bar{\eta}+j\kappa(\bar{\eta})}{\tau_o} \leq -\frac{\pi}{2} \end{cases} \tag{3.4.17}$$

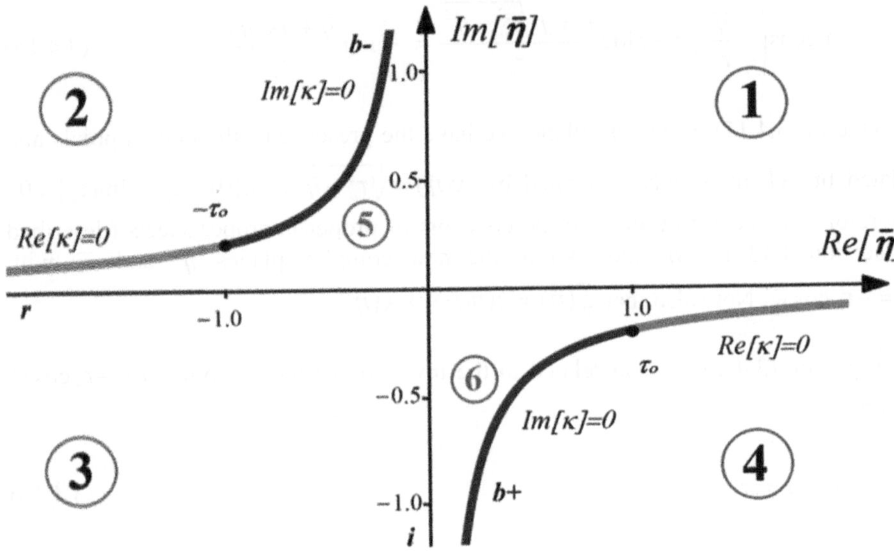

Figure 3.4.1. *Proper sheet of the $\bar{\eta}$ -plane with four quadrants 1, 2, 3 and 4 delimited by the images of the standard branch lines labeled with b+ and b- (the branches with $Im[\kappa(\bar{\eta})] = 0$ are assumed as standard). The proper sheet is defined on the constraint $Im[\kappa(\bar{\eta})] < 0$. Sub-regions 5 and 6 are respectively part of the second and fourth quadrants of the $\bar{\eta}$ -plane; however, they show a different sign on the real part of κ with respect to the rest of their quadrant. In this figure, τ_o is equal to 1-j0.2. For a color version of the figure, see www.iste.co.uk/daniele/ scattering1.zip*

In this context, the images of the proper and improper sheets of the $\bar{\eta}$ -plane in the $\bar{w}$ -plane are presented in Figure 3.4.2. The images of the proper and improper sheets of the $\bar{\eta}$ -plane in the $\bar{w}$ -plane are with gray and blue colors, respectively, in Figure 3.4.2. Note that the proper and improper sheets of $\bar{\eta}$ are replicated periodically in the $\bar{w}$ -plane with period 2π. With alternative assumptions of branch lines, the images of the proper and improper sheets are different from the one described in Figure 3.4.2 (see also discussion in section 1.13).

In particular, for real τ_o, the image of the principal plane is the strip $S_p : -\pi \le \mathrm{Re}[\bar{w}] \le 0$ that corresponds to $S_p : -\Phi \le \mathrm{Re}[w] \le 0$. We define S_p as the principal (proper) strip.

Recalling from Chapter 1 that when the WH equations involve multivalued functions, we preliminary limit the investigation to the study of solutions only in the fundamental sheet. Taking into account that we assume the proper sheet as the fundamental sheet, we consider the singularities of $\bar{X}_+(\bar{\eta})$ and $\bar{X}_-(\bar{\eta})$ only located in the proper sheet of the $\bar{\eta}$ -plane.

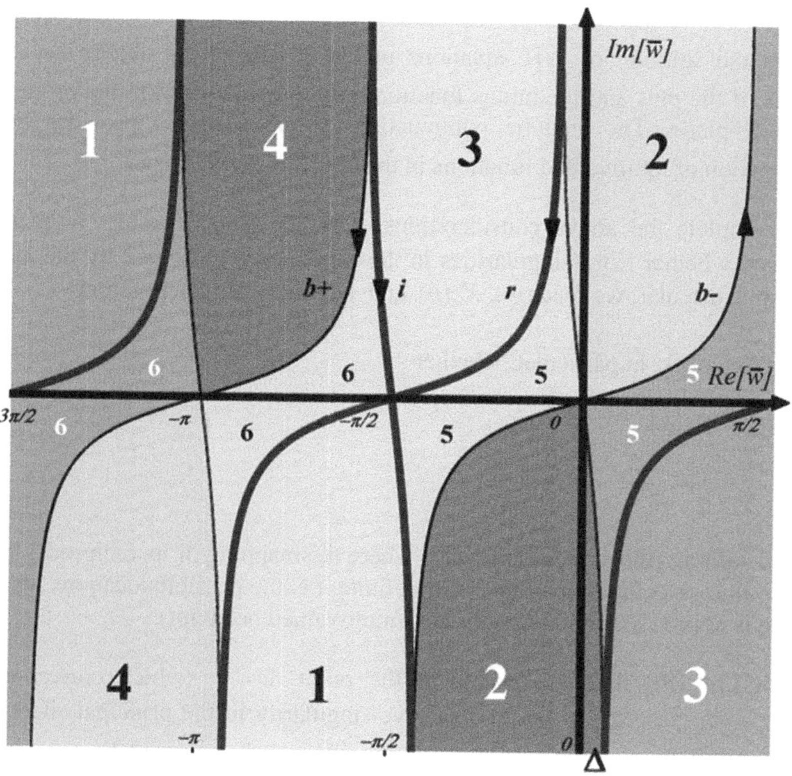

Figure 3.4.2. *Images of the proper and improper sheets of the $\bar{\eta}$ -plane in the $\bar{w}$ -plane are respectively shown with gray and blue colors. For the proper sheet, the images of four quadrants 1, 2, 3 and 4 are delimited by the images of the standard branch lines b+ and b-. The r and i lines are the images of the real and imaginary axes of the $\bar{\eta}$ -plane. Sub-regions 5 and 6 are respectively part of the second and fourth quadrants of the $\bar{\eta}$ -plane, but they show a different sign on the real part of κ with respect to the rest of their quadrants. Δ is $arctan(-Im[\tau_o]/Re[\tau_o])$. Here, τ_o is equal to 1-j0.2. The proper and improper sheets of the $\bar{\eta}$ -plane are replicated periodically in the $\bar{w}$ -plane with period 2π. For a color version of the figure, see www.iste.co.uk/daniele/scattering1.zip*

For example, let us study the branch point $\eta = +\tau_o$ of the η -plane in the $\bar{\eta}$ -plane, which originates from the ξ function. According to (3.4.1), this point corresponds to $\bar{\eta} = -\tau_o \cos\frac{\pi^2}{\Phi}$ (also $w = -\pi \Rightarrow \bar{w} = -\pi^2/\Phi$). We observe that

$\bar{\eta}=-\tau_o\cos\frac{\pi^2}{\Phi}$ ($\bar{w}=-\pi^2/\Phi$) is not located in the principal sheet, which is $S_p:-\pi\le \mathrm{Re}[\bar{w}]\le 0$.

From the solution of WH equations in the $\bar{\eta}$-plane, we obtain the analytic elements of the plus and the minus functions that are only valid in the proper sheet of the $\bar{\eta}$-plane. The analytic continuation of the solutions provides the full representation of multivalued functions in the entire complex planes.

To complete the above considerations, with the help of using $\bar{w}$-plane, we investigate whether extra singularities in the η-plane are generated by the mapping (3.4.1). In particular, we focus on $X_+(\eta)$ and $Y_+(-m)=X_-(m)$ functions.

First we check, in particular, whether

$$\frac{d\bar{\eta}}{d\eta}=\frac{\sin\bar{w}}{\sin\frac{\Phi}{\pi}\bar{w}} \qquad (3.4.18)$$

is null or infinite (this condition reveals where the mapping or its reciprocal have an infinite value, thus the derivative is not definite, i.e. the possible locations where the mapping is not analytic and possibly with multivalued behavior).

From (3.4.18), first, we nominate the point $\bar{w}=0$ (which corresponds to $w=0,\eta=-\tau_o,\bar{\eta}=-\tau_o$) to be an offensive singularity in the principal sheet of the $\bar{\eta}$-plane at $\bar{\eta}=-\tau_o$. However, this singularity is only apparent because in w=0, $\left.\frac{d\bar{\eta}}{d\eta}\right|_{w=0}=\frac{\pi}{\Phi}$. This means that the plus function $X_+(\eta)$ regular in $\eta=-\tau_o(w=0)$ is mapped onto $\bar{X}_+(\bar{\eta})$ that remains regular at $\bar{\eta}=-\tau_o\cos\bar{w}\big|_{\bar{w}=0}=-\tau_o$.

Conversely, the point $\bar{w}=-\pi$ (which corresponds to $w=-\Phi,\eta=-\tau_o\cos\Phi,\bar{\eta}=\tau_o$) generates a singularity related to the mapping (3.4.1). Hence, although $X_+(\eta)$ is regular at $\eta=-\tau_o\cos\Phi$, $\bar{X}_+(\bar{\eta})$ has a branch point at $\bar{\eta}=\tau_o$.

We observe that the branch point $\bar{\eta} = \tau_o$ is not offensive in the WH procedure, since it does not compromise the regularity properties of plus functions defined in the $\bar{\eta}$-plane.

Concerning minus functions, we observe that the mapping $m \to \eta \to \bar{\eta}$ (3.4.7) introduces possible singularities in the $\bar{\eta}$-plane where

$$\frac{d\bar{\eta}}{dm} = \frac{\sin \bar{w}}{\sin \frac{\Phi}{\pi}(\bar{w}+\pi)} \tag{3.4.19}$$

is null or infinite.

From (3.4.19), first, we nominate the point $\bar{w} = -\pi$ (which corresponds to $w = -\Phi, \eta = -\tau_o \cos\Phi, \bar{\eta} = \tau_o, m = \tau_o$) to be an offensive singularity in the principal sheet of the $\bar{\eta}$-plane at $\bar{\eta} = \tau_o$. However, this singularity is only apparent because in $\bar{w} = -\pi$, $\left.\frac{d\bar{\eta}}{dm}\right|_{w=-\pi} = -\frac{\pi}{\Phi}$. This means that the minus function $X_-(m)$ regular in $m = \tau_o$ ($w = -\Phi$) is mapped onto $\bar{X}_-(\bar{\eta})$ that remains regular at $\bar{\eta} = -\tau_o \cos \bar{w}\big|_{\bar{w}=-\pi} = \tau_o$.

Conversely, the point $\bar{w} = 0$ (which corresponds to $w = 0, \eta = -\tau_o, \bar{\eta} = -\tau_o, m = \tau_o \cos\Phi$) generates a singularity related to the mapping (3.4.1). Hence, although $X_-(m)$ is regular at $m = \tau_o \cos\Phi$, $\bar{X}_-(\bar{\eta})$ has a branch point at $\bar{\eta} = -\tau_o$.

We observe that the branch point $\bar{\eta} = -\tau_o$ is not offensive in the WH procedure, since it does not compromise the regularity properties of minus functions that are also defined in the $\bar{\eta}$-plane.

To anticipate possible singularities present in $\bar{X}_+(\bar{\eta})$ that arise from the singularities of $X_+(\eta)$, we take into account: a) branch points $\eta = \pm\tau_o$, b) a finite number of poles (structural surface waves) and c) a limited number of poles that arise from the GO fields (the source):

a) the point τ_o is a branch point for $F_+(\eta)$. It yields $w = -\pi$ or $\bar{w} = -\frac{\pi^2}{\Phi}$ that lies out of the proper $\bar{\eta}$-plane (Figure 3.4.2). Similarly, the point $-\tau_o$ is a branch point for $F_-(m)$. It yields $m = \tau_o \cos(w + \Phi) = -\tau_o$ or $w = -\Phi + \pi$, $\bar{w} = -\pi + \frac{\pi^2}{\Phi} > 0$ that again lies out the proper sheet of the $\bar{\eta}$ -plane (Figure 3.4.2);

b) the surface waves are characterized by factors $e^{-j\chi_{sur}\rho}$ (while observing $\varphi = \Phi$) where, for vanishing losses, χ_{sur} is real: $\chi_{sur} > \tau_o$. In the w plane, it yields $\chi_{sur} = -\tau_o \cos w_{sur} \rightarrow w_{sur} = -\pi + jw''_{sur} \rightarrow \bar{w}_{sur} \rightarrow -\frac{\pi}{\Phi} + j\frac{w''_{sur}}{\Phi}$. From Figure 3.4.2, it means that the surface wave poles are out the proper sheet, hence the surface waves are not to be considered in the spectra $\bar{X}_+(\bar{\eta})$ while considering the WH equations in the principal sheet. However, the occurrence of these poles is important when we evaluate the field for an arbitrary direction φ (recall that $X_+(\eta)$ and $X_-(m)$ are spectra defined respectively for $\varphi = 0, \Phi$). In fact, according to section 3.10, the spectra in the w plane for an arbitrary direction φ is given by (3.10.5) in terms of $\hat{X}_+(w \pm \varphi)$ that require evaluations in the improper sheets. As a consequence, asymptotic estimations of fields can capture potential surface wave poles (see examples in Chapter 5);

c) in the solution of the WH equations, the non-standard GO poles located in the proper (fundamental) sheet play a fundamental role. A complete description of the GO contributions can create some difficulty in the presence of an acute aperture wedge ($\Phi < \pi / 2$) where several internal reflections are excited. However, in the context of the electromagnetic scattering problem by impenetrable wedges, formulated in terms of equation (3.5.10), the evaluation of non-standard GO contributions is simple (see section 3.7).

In summary, the possibility of dealing with plus and minus functions defined in the same complex plane is very important. In particular, we now have the possibility of eliminating the minus functions in the generalized Wiener-Hopf equations (GWHEs), that arise from the functional equations, by using simple applications of Cauchy decomposition formulas given in Chapter 1. This property forms the basis of the Fredholm factorization method (see simple applications in Chapter 1) that the authors have recently developed to solve complex diffraction problems involving angular and planar regions (Daniele and Lombardi 2016; Daniele *et al.* 2017a, 2017b, 2018).

Furthermore, when available, the classical WH factorization of matrix kernels can be applied to solve GWHEs after the application of the mapping (3.4.1).

3.5. Generalized Wiener-Hopf equations for the impenetrable wedge scattering problem

From the generalized Wiener-Hopf functional equations of four angular regions given in (3.3.57)–(3.3.64), in this section, we consider the electromagnetic scattering of an impenetrable wedge modeled by anisotropic surface impedances illuminated by a plane wave.

According to Figure 3.3.2, in order to simplify the procedure and without loss of generality, we consider that the impenetrable wedge consists of regions 3 and 4, and regions 1 and 2 have the same aperture angle, i.e. $\gamma_a = \Phi$ and $\gamma_b = \Phi$ (which is always possible using an appropriate orientation).

These assumptions are convenient since they allow us to use the same mapping (3.4.1) in functional equations (3.3.57)–(3.3.60) for regions 1 and 2.

We assume Leontovich boundary conditions on the faces of the wedge $\varphi = \Phi$ and $\varphi = -\Phi$ in terms of matrix impedances (Senior and Volakis 1995):

$$\begin{bmatrix} E_z(\rho,\Phi) \\ E_\rho(\rho,\Phi) \end{bmatrix} = Z_a \begin{bmatrix} H_\rho(\rho,\Phi) \\ -H_z(\rho,\Phi) \end{bmatrix}, \begin{bmatrix} E_z(\rho,-\Phi) \\ E_\rho(\rho,-\Phi) \end{bmatrix} = -Z_b \begin{bmatrix} H_\rho(\rho,-\Phi) \\ -H_z(\rho,-\Phi) \end{bmatrix} \tag{3.5.1}$$

with the matrices $Z_{a,b} = Z_o \begin{bmatrix} z_{11}^{a,b} & z_{12}^{a,b} \\ z_{21}^{a,b} & z_{22}^{a,b} \end{bmatrix}$ depending on the wedge material.

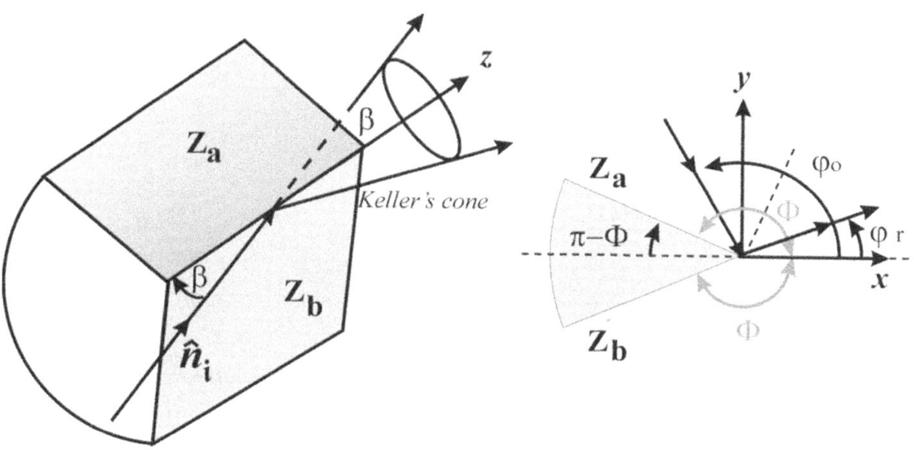

Figure 3.5.1. *Scattering of an impenetrable wedge by a plane wave at skew incidence. For a color version of the figure, see www.iste.co.uk/daniele/scattering1.zip*

Figure 3.5.1 presents the scattering problem of an impenetrable wedge by a plane wave at skew incidence. In particular, the structure is illuminated by a plane wave described in (3.3.66).

By applying boundary conditions (3.5.1) to the functional equations (3.3.57)–(3.3.60), we obtain the system of four generalized Wiener-Hopf equations (GWHEs):

$$S(\eta)X_+(\eta) = D(\eta)X_-(m) \tag{3.5.2}$$

where:

$$X_+(\eta) = \begin{vmatrix} V_{z+}(\eta,0) \\ V_{\rho+}(\eta,0) \\ Z_o I_{z+}(\eta,0) \\ Z_o I_{\rho+}(\eta,0) \end{vmatrix}, \qquad X_-(m) = \begin{vmatrix} Z_o I_{\rho+}(-m,\Phi) \\ -Z_o I_{z+}(-m,\Phi) \\ -Z_o I_{\rho+}(-m,-\Phi) \\ Z_o I_{z+}(-m,-\Phi) \end{vmatrix} \tag{3.5.3}$$

$$S(\eta) = \begin{vmatrix} \xi(\eta) & 0 & -\dfrac{\alpha_o\eta}{k} & -\dfrac{\tau_o^2}{k} \\ \dfrac{\alpha_o\eta}{k} & \dfrac{\tau_o^2}{k} & \xi(\eta) & 0 \\ \xi(\eta) & 0 & \dfrac{\alpha_o\eta}{k} & \dfrac{\tau_o^2}{k} \\ -\dfrac{\alpha_o\eta}{k} & -\dfrac{\tau_o^2}{k} & \xi(\eta) & 0 \end{vmatrix} \tag{3.5.4}$$

$$D(\eta) =$$

$$\begin{vmatrix} -n(\eta)\,z_{11}^a - \dfrac{\tau_o^2}{k} & -n(\eta)\,z_{12}^a - \dfrac{m(\eta)\alpha_o}{k} & 0 & 0 \\ -\dfrac{m(\eta)\,z_{11}^a\alpha_o}{k} + \dfrac{z_{21}^a\tau_o^2}{k} & n(\eta) - \dfrac{m(\eta)\,z_{12}^a\alpha_o}{k} + \dfrac{z_{22}^a\tau_o^2}{k} & 0 & 0 \\ 0 & 0 & -n(\eta)\,z_{11}^b - \dfrac{\tau_o^2}{k} & -n(\eta)\,z_{12}^b - \dfrac{m(\eta)\alpha_o}{k} \\ 0 & 0 & \dfrac{m(\eta)\,z_{11}^b\alpha_o}{k} - \dfrac{z_{21}^b\tau_o^2}{k} & -n(\eta) + \dfrac{m(\eta)\,z_{12}^b\alpha_o}{k} - \dfrac{z_{22}^b\tau_o^2}{k} \end{vmatrix} \tag{3.5.5}$$

where:

$$m(\eta) = -\eta \cos\Phi + \xi(\eta)\sin\Phi \tag{3.5.6}$$

$$n(\eta) = -\eta \sin\Phi - \xi(\eta)\cos\Phi \tag{3.5.7}$$

with the spectral propagation constant along y $\xi(\eta) = \sqrt{\tau_o^2 - \eta^2}$. According to Figure 3.3.1, the Laplace transforms (3.5.3) are defined as $F_+(\eta, \varphi = 0) = \int_0^\infty f_+(\rho, \varphi = 0) e^{+j\eta\rho} d\rho$, with $\rho = x$, and $F_+(-m, \varphi = \pm\Phi) = \int_0^\infty f_-(\rho, \varphi = \pm\Phi)$ $e^{-jm\rho} d\rho$, with $\rho = v$.

Note that (3.5.2) are generalized equations since plus and minus unknowns ($X_+(\eta)$, $X_-(m)$) are defined in different complex planes η and m, related together via (3.5.6).

In normal form, (3.5.2) becomes

$$G(\eta) X_+(\eta) = X_-(m) \tag{3.5.8}$$

where the matrix WH kernel is

$$G(\eta) = D^{-1}(\eta) S(\eta) \tag{3.5.9}$$

By directly applying the mapping (3.4.1) to (3.5.9) (the same for the two angular regions 1 and 2), we obtain a system of classical Wiener-Hopf equations (CWHEs):

$$\bar{G}(\bar{\eta}) \bar{X}_+(\bar{\eta}) = \bar{X}_-(\bar{\eta}) \tag{3.5.10}$$

where the matrix WH kernel is

$$\bar{G}(\bar{\eta}) = D^{-1}(\eta(\bar{\eta})) S(\eta(\bar{\eta})) \tag{3.5.11}$$

3.6. Solution of generalized Wiener-Hopf equations for the impenetrable wedge scattering problem

We presented two ways for solving WH equations in Chapter 1: the classical method of factorization and the Fredholm factorization method.

By applying the classical method of factorization, described in section 1.5 (1.5.13) and (1.5.14), to (3.5.10), it gives the solution

$$\overline{X}_+(\overline{\eta}) = \sum_i \frac{\left[\overline{G}_+(\overline{\eta})\right]^{-1} \overline{G}_+(\overline{\eta}_i) R_{\overline{X}_+}^{n.s.}(\overline{\eta}_i)}{\overline{\eta} - \overline{\eta}_i} + \sum_i \frac{\left[\overline{G}_+(\overline{\eta})\right]^{-1} \overline{G}_-^{-1}(\overline{\eta}_i) R_{\overline{X}_-}^{n.s.}(\overline{\eta}_i)}{\overline{\eta} - \overline{\eta}_i} \tag{3.6.1}$$

$$\overline{X}_-(\overline{\eta}) = \overline{G}(\overline{\eta}) \overline{X}_+(\overline{\eta}) \tag{3.6.2}$$

if the factorization of the kernel $\overline{G}(\overline{\eta})$ is available in closed form.

By applying the Fredholm factorization method, described in section 1.9, to (3.5.10), it yields

$$\overline{X}_+(\overline{\eta}) + \frac{\left[\overline{G}(\overline{\eta})\right]^{-1}}{2\pi j} \int_{-\infty}^{\infty} \frac{\left[\overline{G}(\overline{\eta}') - \overline{G}(\overline{\eta})\right] \overline{X}_+(\overline{\eta})}{\overline{\eta}' - \overline{\eta}} d\overline{\eta}' = \left[\overline{G}(\overline{\eta})\right]^{-1} \sum_i \frac{R_{\overline{X}_-}^{n.s.}(\overline{\eta}_i)}{\overline{\eta} - \overline{\eta}_i} + \sum_i \frac{R_{\overline{X}_+}^{n.s.}(\overline{\eta}_i)}{\overline{\eta} - \overline{\eta}_i} \tag{3.6.3}$$

for $\mathrm{Im}[\overline{\eta}] = 0$ and (3.6.2). The Fredholm factorization method can be directly applied to the angular complex plane $\overline{w}$ following the properties described in section 1.15. Numerical demonstrations and validations of the approximate solution (3.6.3) are presented in Chapter 5. In particular, here we anticipate that the solution of isolated wedge problems with the Fredholm factorization method is particularly effective in the $\overline{w}$ -plane due to the properties of the kernel. Moreover, the Fredholm factorization method can always be applied to obtain the solution of the problem.

While in this chapter we provide further details on the classical solution (3.6.1), Chapter 5 is completely devoted to the comprehensive presentation of the Fredholm factorization method for the solution of impenetrable wedge scattering problems, modeled by (3.5.10).

Both solutions (3.6.1) and (3.6.3) require the non-standard part of the plus and minus functions, which is the topic of the following section.

3.7. Non-standard parts of the plus and minus functions in GWHEs for the impenetrable wedge scattering problem

We recall that the non-standard parts of the plus and minus functions arise from the GO poles respectively located in $\mathrm{Im}[\overline{\eta}] > 0$ and $\mathrm{Im}[\overline{\eta}] < 0$, i.e. in the regularity half-plane.

According to Figure 3.5.1, in general, a plane wave with a unit vector $\hat{\mathbf{n}}_i = -\tau_{oi} \cos\varphi_i \hat{\mathbf{x}} - \tau_{oi} \sin\varphi_i \hat{\mathbf{y}} + \alpha_{oi} \hat{\mathbf{z}}$ produces a phase factor:

$$e^{-j\hat{\mathbf{n}}_i \cdot \mathbf{r}} = e^{j\tau_{oi}\rho\cos(\varphi-\varphi_i)} e^{-j\alpha_{oi} z} \tag{3.7.1}$$

where $\alpha_o = k\cos\beta_i$ and $\tau_o = k\sin\beta_i$ with the skewness incident zenithal angle $\vartheta = \beta_i$ and the azimuthal angle φ_i.

Let us assume a plane wave with the incidence angle φ_n (incoming) impinging on the face "a": $\varphi = +\Phi$. According to the boundary conditions, the reflected wave presents an (incoming) impinging angle φ_{n+1} such that $\cos(\Phi - \varphi_n) = \cos(\Phi - \varphi_{n+1})$ or $\varphi_{n+1} = -\varphi_n + 2\Phi$ (Snell law).

In order to evaluate the outgoing angle, it is sufficient to sum $-\pi$.

For the face b ($\varphi = -\Phi$), the Snell law becomes $\varphi_{n+1} = -\varphi_n - 2\Phi$ for incoming waves.

By indicating with φ_o the incidence angle of a plane wave that excites the wedge, a set of possible multiple reflected waves is given by the incoming angles $\varphi_n = -\varphi_o + 2n\Phi$ ($n = 0, \pm 1, \pm 2, \ldots$) (where n=0 is the incident wave).

Let us consider the direction $\varphi = 0$. By applying the Laplace transforms (3.3.2) and (3.3.3), we observe that each plane wave of the aforementioned set yields different poles $\eta_n = -\tau_o \cos\varphi_n$ in the η-plane, while in the $\bar{\eta}$-plane we obtain the corresponding poles at the same location $\bar{\eta}_n = -\tau_o \cos\frac{\pi}{\Phi}\varphi_n = -\tau_o \cos\frac{\pi}{\Phi}\varphi_o = \bar{\eta}_o$ but for different n in different sheets of the complex plane $\bar{\eta}$.

In fact, each plane wave presents different locations in the $\bar{w}$ and w planes where they assume the values

$$\bar{w}_n = \pm\frac{\pi}{\Phi}\varphi_n = \mp\bar{w}_o + 2n\pi\,, \quad n = 0, \pm 1, \pm 2, \ldots \tag{3.7.2}$$

where $\bar{w}_o = -\frac{\pi}{\Phi}\varphi_o$.

We recall that the solution of the WH equation is first obtained in the proper sheet; therefore, we only consider the subset of GO poles $\bar{\eta}_n$ located in the proper sheet of the complex plane $\bar{\eta}$, in particular considering their position in the angular complex plane $\bar{w}$ (see Figure 3.4.2).

Without loss of generality, we assume that $0 \le \varphi_o \le \Phi$. From Figure 3.4.2, it follows that only one pole is located in the proper sheet, which is the one corresponding to φ_o, i.e. $w_o = -\varphi_o$, $\eta_o = -\tau_o \cos\varphi_o$, $\bar{w}_o = -\frac{\pi}{\Phi}\varphi_o$, $\bar{\eta}_o = -\tau_o \cos\frac{\pi}{\Phi}\varphi_o$.

Although we have only the pole $\bar{\eta}_o$ in the proper sheet while considering the spectra for $\varphi = 0$, i.e. for $\bar{X}_+(\bar{\eta})$, in general, the corresponding residue can be related to incident and reflected waves from the impedance faces a) $\varphi = \Phi$ and b) $\varphi = -\Phi$. This is the case for non-standard minus function $\bar{X}_-(\bar{\eta})$ that is related to the spectra for $\varphi = \pm\Phi$.

We also recall that only the non-standard part of the GO contributions located in the proper sheet must be considered in the solution procedure, outlined at the end of the previous section.

We now present a remarkable property. In order to solve the GWHEs of the scattering by an impenetrable wedge, it is not necessary to estimate the reflection coefficients on the impedance faces a and b. In the following, we will show that we only need the computation of spectra of the incident wave at $\varphi = 0$.

Let us consider the GO spectra $\bar{X}_+^{GO}(\bar{\eta})$, i.e. the one in the direction $\varphi = 0$. The non-standard contribution occurs for $\mathrm{Im}[\bar{\eta}_o] = \mathrm{Im}[-\tau_o \cos(\pi\varphi_o / \Phi)] > 0$, i.e. $\varphi_o < \Phi / 2$. At the same time, with the condition $\varphi_o < \Phi / 2$, we have that $\bar{X}_-^{GO}(\bar{\eta})$ is standard. In particular, the n.s. contribution $\bar{X}_+^{GOn.s}(\bar{\eta})$ only arises from the incident wave.

Taking into account the expression of the incident plane wave (3.3.66) and the Laplace transforms (3.3.2) and (3.3.3), we obtain:

$$X_+^i(\eta) = \frac{T_o}{\eta + \tau_o \cos\varphi_o} \tag{3.7.3}$$

with

$$T_o = \begin{vmatrix} jE_o \\ j\dfrac{\alpha_o \cos\phi_o E_o + kZ_o \sin\phi_o H_o}{\tau_o} \\ jZ_o H_o \\ j\dfrac{\alpha_o Z_o \cos\phi_o H_o - k\sin\phi_o E_o}{\tau_o} \end{vmatrix} \tag{3.7.4}$$

It follows the evaluation of the residue in the $\bar{\eta}$ -plane from that obtained in the η -plane:

$$\bar{T}_o = T_o \left.\frac{d\bar{\eta}}{d\eta}\right|_{w=-\varphi_o} = T_o \frac{\pi}{\Phi} \frac{\sin\frac{\pi}{\Phi}\varphi_o}{\sin\varphi_o}$$

thus in (3.6.1) and (3.6.3) we have

$$R_{\bar{X}_+}^{n.s.}(\bar{\eta}_o) = \bar{T}_o u(\frac{\Phi}{2} - \varphi_o) \tag{3.7.5}$$

where u(x) is the Heaviside step function.

Using (3.7.5) in (3.6.1), we obtain the classical solution

$$\bar{X}_+(\bar{\eta}) = \frac{\left[\bar{G}_+(\bar{\eta})\right]^{-1} \bar{G}_+(\bar{\eta}_o)\bar{T}_o}{\bar{\eta} - \bar{\eta}_o} u(\frac{\Phi}{2} - \varphi_o) + \frac{\left[\bar{G}_+(\bar{\eta})\right]^{-1} \bar{G}_-^{-1}(\bar{\eta}_o) R_{\bar{X}_-}^{n.s.}(\bar{\eta}_o)}{\bar{\eta} - \bar{\eta}_o} \tag{3.7.6}$$

$$\bar{X}_-(\bar{\eta}) = \bar{G}(\bar{\eta})\bar{X}_+(\bar{\eta}) \tag{3.7.7}$$

Conversely, if $\varphi_o > \Phi/2$, $\bar{X}_+^{GO}(\bar{\eta})$ is standard while $\bar{X}_-^{GO}(\bar{\eta})$ is non-standard.

Using (3.7.7), we can immediately estimate the residue $R_{\bar{X}_-}^{n.s.}$ at $\bar{\eta}_o = -\tau_o \cos\frac{\pi}{\Phi}\varphi_o$ in the form

$$R_{\bar{X}_-}^{n.s.}(\bar{\eta}_o) = \bar{G}(\bar{\eta}_o) R_{\bar{X}_+}^{s.}(\bar{\eta}_o) = \bar{G}(\bar{\eta}_o)\bar{T}_o u(\varphi_o - \frac{\Phi}{2}) \tag{3.7.8}$$

since only the incident wave contribution has poles in the proper sheet of $\bar{\eta}$ for the plus function $\bar{X}_+(\bar{\eta})$. In this case, $\bar{X}_+(\bar{\eta})$ is standard, with residue in $\bar{\eta}=\bar{\eta}_o$ $R^{s_i}_{\bar{X}_+}(\bar{\eta}_o)=\bar{T}_o u(\varphi_o-\frac{\Phi}{2})$ where $\bar{T}_o$ is described in (3.7.5). Most importantly, the knowledge of $\bar{G}(\bar{\eta}_o)$ avoids the need of a direct evaluation of the reflection coefficients/components for the computation of the non-standard part of the minus function $\bar{X}_-(\bar{\eta})$.

In summary, once the factorization of the matrix kernel $\bar{G}(\bar{\eta})=\bar{G}_-(\bar{\eta})\bar{G}_+(\bar{\eta})$ is provided, the classical solution of the impenetrable wedge is given by substituting (3.7.8) into (3.7.6):

$$\bar{X}_+(\bar{\eta})=\frac{\left[\bar{G}_+(\bar{\eta})\right]^{-1}\bar{G}_+(\bar{\eta}_o)\bar{T}_o}{\bar{\eta}-\bar{\eta}_o} \tag{3.7.9}$$

with $\bar{T}_o$ described in (3.7.4) and (3.7.5) for both standard and non-standard $\bar{X}_+(\bar{\eta})$.

3.8. Closed-form solution of the PEC wedge scattering problem at normal incidence

Taking into account that $Z^a=Z^b=0$, $\alpha_o=0$, equation (3.5.8) reduces to a simple system of order 2:

$$G(\eta)F_+(\eta)=F_-(m) \tag{3.8.1}$$

where:

$$G(\eta)=\begin{vmatrix}\xi & -kZ_o\\ \xi & kZ_o\end{vmatrix},\quad F_+(\eta)=\begin{vmatrix}V_{z+}(\eta,0)\\ I_{\rho+}(\eta,0)\end{vmatrix},\quad F_-(m)=\begin{vmatrix}-kZ_oI_{\rho+}(-m,\Phi)\\ kZ_oI_{\rho+}(-m,-\Phi)\end{vmatrix} \tag{3.8.2}$$

By summing the two equations of (3.8.1), we obtain the scalar GWHE:

$$\xi\, X_+(\eta)=Y_+(-m) \tag{3.8.3}$$

with $X_+(\eta)=2V_{z+}(\eta,0)$, $Y_+(-m)=kZ_o\,[I_{\rho+}(m,-\Phi)-I_{\rho+}(m,\Phi)]$.

The mapping $\eta = \eta(\bar{\eta})$ (3.4.1) yields the following scalar CWHE in the $\bar{\eta}$ -plane:

$$\bar{\xi}(\bar{\eta}))\bar{X}_+(\bar{\eta}) = \bar{Y}_+(-\bar{\eta}) = \bar{X}\_(\bar{\eta}) \tag{3.8.4}$$

The solution of (3.8.4) given by (3.7.9) in explicit form is

$$\bar{X}_+(\bar{\eta}) = \frac{\left[\bar{\xi}_+(\bar{\eta})\right]^{-1} \bar{\xi}_+(\bar{\eta}_o)\bar{T}_o}{\bar{\eta} - \bar{\eta}_o} \tag{3.8.5}$$

where $\bar{T}_o$, described in (3.7.5) and (3.7.6), is the non-standard part of $X_+(\eta) = 2V_{z+}(\eta, 0)$:

$$\bar{T}_o = 2jE_o \frac{\pi}{\Phi} \frac{\sin\frac{\pi}{\Phi}\varphi_o}{\sin\varphi_o} \tag{3.8.6}$$

since here

$$X_+^i(\eta) = \frac{T_o}{\eta + k\cos\varphi_o} \tag{3.8.7}$$

with $T_o = j2E_o$.

The key point to solve the CWHE is the factorization of the kernel $\bar{\xi}(\bar{\eta})) = \bar{\xi}_-(\bar{\eta}))\bar{\xi}_+(\bar{\eta}))$. Since we deal with scalar functions, this factorization can be accomplished by exploiting the properties of plus and minus functions in the angular complex plane $\bar{w}$, as described in section 3.4.

Since

$$\xi = \sqrt{\tau_o{}^2 - \eta^2} = -\tau_o \sin w = (-\sqrt{\tau_o}\sin\frac{1}{2}\bar{w})\frac{\sqrt{\tau_o}\sin\frac{\Phi}{\pi}\bar{w}}{\sin\frac{1}{2}\bar{w}} \tag{3.8.8}$$

we define by inspection

$$\bar{\xi}_-(\bar{\eta}) = \sqrt{\tau_o}\sin\frac{1}{2}\bar{w} = \sqrt{\frac{\tau_o + \bar{\eta}}{2}} \tag{3.8.9}$$

$$\bar{\xi}_+(\bar{\eta}) = -\frac{\sqrt{\tau_o}\,\sin\frac{\Phi}{\pi}\bar{w}}{\sin\frac{1}{2}\bar{w}} = \frac{\xi(\eta(\bar{\eta}))}{\bar{\xi}_-(\bar{\eta})} \tag{3.8.10}$$

Equation (3.8.9) is a minus function since it does not change when $\bar{w}$ is substituted by $-2\pi-\bar{w}$ and is regular at $\bar{w}=-\pi$, while (3.8.10) is a plus function because it is even in $\bar{w}$ and regular in $\bar{w}=0$. Equations (3.8.9) and (3.8.10) also give the representation of $\bar{\xi}_\pm$ in the $\bar{\eta}$-plane ($\bar{\eta}=-k\cos\bar{w}$).

The solution in the $\bar{\eta}$-plane ($V_{z+}(\eta,0)=\bar{X}_+(\bar{\eta})/2$) is explicitly obtained by substituting (3.8.6)–(3.8.10) into (3.8.5).

For the practical computation of field components in the physical domain via the inverse Laplace transform and asymptotics, we finally transform the solution (3.8.5) into the w plane using the mappings of section 3.4:

$$V_{z+}(-k\cos w,0) = 2j\pi\frac{E_o\cos\left(\frac{\pi}{2\Phi}\varphi_o\right)\sin\left(\frac{\pi}{2\Phi}w\right)}{k\Phi\sin w\left[-\cos\left(\frac{\pi}{\Phi}w\right)+\cos\left(\frac{\pi}{\Phi}\varphi_o\right)\right]} =$$

$$V_{z+}(-k\cos w,0) = \frac{jE_o\cos(\varphi_o/n_\Phi)}{n_\Phi\, k\sin w}\left(\frac{1}{\sin w/n_\Phi-\sin\varphi_o/n_\Phi}+\frac{1}{\sin w/n_\Phi+\sin\varphi_o/n_\Phi}\right) \tag{3.8.11}$$

with $n_\Phi = 2\Phi/\pi$.

3.9. Alternative solution of the PEC wedge scattering problem at normal incidence via difference equations

Difference equations can be directly obtained from the GWHEs by using the evenness of the plus functions in the $\bar{w}$- or w-plane. For example, we consider equation (3.8.1) with the matrix WH kernel $G(\eta)$ (3.8.2), or, in particular, (3.8.4)

$$\bar{\xi}(\bar{\eta}))\bar{X}_+(\bar{\eta}) = \bar{Y}_+(-\bar{\eta}) = \bar{X}_-(\bar{\eta}) \tag{3.9.1}$$

that, according to section 3.4, in the w plane becomes

$$\hat{\xi}(w)\hat{X}_+(w) = \hat{Y}_+(w+\Phi) \tag{3.9.2}$$

or equivalently

$$\hat{\xi}(w-\Phi)\hat{X}_+(w-\Phi) = \hat{Y}_+(w) \tag{3.9.3}$$

where the kernel ξ of (3.8.3) becomes $\hat{\xi}(w) = -k\sin(w)$ in the w plane.

We can eliminate $\hat{Y}_+(w)$ by taking into account its evenness property, i.e. $\hat{Y}_+(-w) = \hat{Y}_+(w)$.

It yields:

$$\hat{\xi}(w-\Phi)\hat{X}_+(w-\Phi) = \hat{\xi}(-w-\Phi)\hat{X}_+(-w-\Phi) = \hat{\xi}(-w-\Phi)\hat{X}_+(w+\Phi) \tag{3.9.4}$$

Replacing w with $w-\Phi$, we obtain

$$\hat{X}_+(w) = \left[\hat{\xi}(-w)\right]^{-1}\hat{\xi}(w-2\Phi)\hat{X}_+(w-2\Phi) \tag{3.9.5}$$

This can be rewritten as

$$\hat{X}_+(w+\Phi) = K_\Phi(w)\hat{X}_+(w-\Phi) \tag{3.9.6}$$

where:

$$K_\Phi(w) = \left[\hat{\xi}(-w-\Phi)\right]^{-1}\hat{\xi}(w-\Phi) = -\frac{\sin(w-\Phi)}{\sin(w+\Phi)} \tag{3.9.7}$$

We note that the same procedure can be directly applied to the original system of WH equations (3.8.1).

Concerning the source, taking into account (3.7.2), we observe that only the pole $w_o = -\varphi_o$ of $\hat{X}_+(w)$ is located in the proper sheet with the residue corresponding to the incident plane wave; therefore, we can decompose $\hat{X}_+(w)$ into scattered (s) and incident (i) components.

This means:

$$\hat{X}_+(w) = \hat{X}_+^s(w) + \hat{X}_+^i(w)$$

where the incident field is given by (3.8.7), which in the w plane for this problem is

$$\hat{X}_+^i(w) = \frac{2jE_o}{-k\cos w + k\cos\varphi_o} \tag{3.9.8}$$

The unknown spectrum $\hat{X}_+^s(w)$ does not present poles in the strip $-\Phi \le \mathrm{Re}[w] \le 0$ and taking into account the evenness of $\hat{X}_+^s(w)$ in the strip $-\Phi \le \mathrm{Re}[w] \le +\Phi$.

In non-homogeneous form, equation (3.9.6) becomes

$$\hat{X}_+^s(w+\Phi) = K_\Phi(w)\hat{X}_+^s(w-\Phi) + N_\Phi(w) \tag{3.9.9}$$

where the known term $N_\Phi(w)$ is given by

$$N_\Phi(w) = K_\Phi(w)\hat{X}_+^i(w-\Phi) - \hat{X}_+^i(w+\Phi) \tag{3.9.10}$$

In explicit form, equation (3.9.9) becomes:

$$\hat{V}_d^s(w+\Phi) + \hat{V}_d^s(w-\Phi) = \frac{jE_o\sin(w-\Phi)}{k\cos(w-\Phi) - k\cos\varphi_o} + \frac{jE_o\sin(w+\Phi)}{k\cos(w+\Phi) - k\cos\varphi_o} \tag{3.9.11}$$

where:

$$\hat{V}_d^s(w) = \sin w\hat{V}_+^s(w) \tag{3.9.12}$$

In order to solve the difference equation (3.9.11), we use the MF method (see Appendix 3.A). This method requires the definition of the MF transform

$$\hat{X}[t] = -j\int_{-j\infty}^{j\infty}\left[\hat{X}(w)\right]\exp[+jtw]dw \tag{3.9.13}$$

and we consider the transport property

$$MF[\hat{X}(w+\varphi)] = e^{-jt\varphi}MF[\hat{X}(w)] \tag{3.9.14}$$

that holds while $\hat{X}(w)$ is without singularities in the strip $0 < \mathrm{Re}[w] < \varphi$ or $\varphi < \mathrm{Re}[w] < 0$ according to the sign of φ.

Taking into account the regularity of $\hat{V}_d^s(w)$ in the strip $-\Phi \le \mathrm{Re}[w] \le +\Phi$, we obtain

$$MF[\hat{V}_d^s(w+\Phi)+\hat{V}_d^s(w-\Phi)] = (e^{-jt\Phi}+e^{jt\Phi})\widehat{V}_d^s(t) = 2\cos(\Phi t)\widehat{V}_d^s(t) \quad (3.9.15)$$

We next consider the MF transform of the second member of (3.9.11). The substitution $w\pm\Phi \to w$ and the residue theorem yields (Appendix 3.A)

$$\begin{aligned} MF[\frac{jE_o(\sin(w-\Phi))}{k\cos(w-\Phi)-k\cos\varphi_o}] &= j\frac{E_o}{k}(2\pi e^{jt(\Phi-\varphi_o)} - je^{+jt\Phi}\widehat{I}(t)) \\ MF[\frac{jE_o(\sin(w+\Phi))}{k\cos(w+\Phi)-k\cos\varphi_o}] &= j\frac{E_o}{k}(-2\pi e^{-jt(\Phi-\varphi_o)} - je^{-jt\Phi}\widehat{I}(t)) \end{aligned} \quad (3.9.16)$$

where:

$$\widehat{I}(t) = \int_{-j\infty}^{j\infty}\left[\frac{\sin(w)}{\cos(w)-\cos\varphi_o}\right]\exp[+jtw]dw \quad (3.9.17)$$

From (3.9.16), it yields:

$$\widehat{V}_d^s(t) = -\frac{E_o}{k}2\pi\frac{\sin(t(\Phi-\varphi_o))}{\cos(t\Phi)} + \frac{E_o}{k}\widehat{I}(t) \quad (3.9.18)$$

Taking into account that:

$$\hat{V}_d^i(w) = \frac{jE_o\sin[w]}{-k\cos w + k\cos\varphi_o} \quad (3.9.19)$$

$$\widehat{V}_d^i(t) = -\frac{jE_o}{k}MF[\frac{\sin(w)}{\cos w-\cos\varphi_o}] = -\frac{E_o}{k}\widehat{I}(t) \quad (3.9.20)$$

we obtain

$$\widehat{V}_d(t) = \widehat{V}_d^i(t) + \widehat{V}_d^s(t) = -\frac{E_o}{k}2\pi\frac{\sin(t(\Phi-\varphi_o))}{\cos(t\Phi)} \quad (3.9.21)$$

After some mathematical manipulations, the inverse MF transform of (3.9.21)

$$\hat{V}_d(w) = -\frac{j}{2\pi}\int_{-j\infty}^{j\infty} -\frac{E_o}{k}2\pi\frac{\sin(t(\Phi-\varphi_o))}{\cos(t\Phi)}e^{-jtw}dt = \int_{-j\infty}^{j\infty}\frac{E_o}{k}2\pi\frac{\sin(t(\Phi-\varphi_o))\sin(tw)}{\cos(t\Phi)}dt \quad (3.9.22)$$

yields the solution

$$\hat{V}_d(w) =$$

$$\frac{E_o}{k} j \int_{-\infty}^{\infty} \frac{\sinh(v(\Phi - \varphi_o))\sinh(wv)}{\cosh(v\Phi)} dv = 2j\pi \frac{E_o \cos\left(\frac{\pi}{2\Phi}\varphi_o\right)\sin\left(\frac{\pi}{2\Phi}w\right)}{k\Phi \sin w\left[-\cos\left(\frac{\pi}{\Phi}w\right)+\cos\left(\frac{\pi}{\Phi}\varphi_o\right)\right]}$$

which is identical to (3.8.11).

3.10. Generalized WH functional equations for angular regions in the w plane

To solve wedge problems in this book using the WH technique, we present mathematical methods that use the spectral plane η resulting from the Laplace transform of the electromagnetic field, together with the spectral plane w.

The use of the plane η is of great importance because it allows us to define the entire problem in a unique spectral plane when different materials coexist. It allows us to apply a unified procedure to solve arbitrary problems not amenable to exact solutions.

However, the presence of multivalued functions sometimes makes some mathematical formulations of the proposed methods obscure in the η-plane.

Hence we also present formulations that take advantage of the w plane, where multivalued functions can be more easily treated. However, in counterpart, we note the limit that w planes are always related to a specific wave number k.

From (3.3.57) and (3.3.58) using the definition given in section 3.4, in the w plane, the functional equations of region 1 with $\gamma_a = \varphi$ become

$$\begin{aligned}
&-\tau_o \hat{V}_{zd}(w,0) - \frac{\tau_o^2}{\omega\varepsilon}\hat{I}_{\rho+}(w,0) + \frac{\alpha_o\tau_o}{\omega\varepsilon\tan w}\hat{I}_{zd}(w,0) \\
&= -\tau_o \hat{V}_{zd}(w+\varphi,\varphi) - \frac{\tau_o^2}{\omega\varepsilon}\hat{I}_{\rho+}(w+\varphi,\varphi) + \frac{\alpha_o\tau_o}{\omega\varepsilon\tan(w+\varphi)}\hat{I}_{zd}(w+\varphi,\varphi)
\end{aligned} \quad (3.10.1)$$

$$
\begin{aligned}
&-\tau_o \hat{I}_{zd}(w,0)+\frac{\tau_o^2}{\omega\mu}\hat{V}_{\rho+}(w,0)-\frac{\alpha_o\tau_o}{\omega\mu\tan w}\hat{V}_{zd}(w,0)\\
&=-\tau_o \hat{I}_{zd}(w+\varphi,\varphi)+\frac{\tau_o^2}{\omega\mu}\hat{V}_{\rho+}(w+\varphi,\varphi)-\frac{\alpha_o\tau_o}{\omega\mu\tan(w+\varphi)}\hat{V}_{zd}(w+\varphi,\varphi)
\end{aligned}
\quad (3.10.2)
$$

where:

$$
\begin{aligned}
&\hat{V}_{zd}(w,\varphi)=\sin wV_{z+}(-\tau_o\cos w,\varphi),\quad \hat{I}_{zd}(w,\varphi)=\sin wI_{z+}(-\tau_o\cos w,\varphi),\\
&\hat{V}_{\rho+}(w,\varphi)=V_{\rho+}(-\tau_o\cos w,\varphi),\quad \hat{I}_{\rho+}(w,\varphi)=I_{\rho+}(-\tau_o\cos w,\varphi)
\end{aligned}
\quad (3.10.3)
$$

Similar expressions hold in the other angular regions.

While the GWHEs were obtained in the η-plane with some limitations regarding the location of η (e.g. since η must lie in the proper sheet, $\gamma_a \leq \pi$), the above equations hold for any value of w as well as for $\varphi > \pi$.

Moreover, the functional equations of the four regions (3.3.57)–(3.3.64) in the w plane yield the possibility of evaluating the spectral field component in w for an arbitrary direction φ.

In particular, the spectra along an arbitrary φ (3.10.3) can be obtained in terms of the axial spectra (i.e. for $\varphi = 0$):

$$
\begin{aligned}
&\hat{V}_{zd}(w,0)=\sin wV_{z+}(-\tau_o\cos w,0),\quad \hat{I}_{zd}(w,0)=\sin wI_{z+}(-\tau_o\cos w,0),\\
&\hat{V}_{\rho+}(w,0)=V_{\rho+}(-\tau_o\cos w,0),\quad \hat{I}_{\rho+}(w,0)=I_{\rho+}(-\tau_o\cos w,0)
\end{aligned}
\quad (3.10.4)
$$

For region 1, we obtain the spectra (3.10.3) with a set of four equations: the first two are obtained from (3.10.1) and (3.10.2) by substituting $w+\varphi$ with w and the second two from (3.10.1) and (3.10.2) by using the following steps: 1) substitute w with –w, 2) apply symmetry (plus functions are even in w and d functions are odd in w) to obtain a positive w and 3) substitute $w-\varphi$ with w.

The set of four equations allows the representation of (3.10.3) in terms of shifted axial spectra (3.10.4):

$$\begin{aligned}
V_{\rho+}(w,\varphi) &= \frac{1}{2k\tau_o}(Z_o(-(k^2+\cot(w)\cot(w-\varphi)\alpha_o^2)I_{zd}(w-\varphi) \\
&+(k^2+\cot(w)\cot(w+\varphi)\alpha_o^2)I_{zd}(w+\varphi)+\cot(w)\alpha_o\tau_o(I_{\rho+}(w-\varphi)-I_{\rho+}(w+\varphi))) \\
&+k(\csc(w)\sin(\varphi)\alpha_o(-\csc(w-\varphi)V_{zd}(w-\varphi)+\csc(w+\varphi)V_{zd}(w+\varphi)) \\
&+\tau_o(V_{\rho+}(w-\varphi)+V_{\rho+}(w+\varphi))))
\end{aligned}$$

$$\begin{aligned}
&V_{zd}(w,\varphi) \\
&=\frac{Z_o(\alpha_o(-\cot(w-\varphi)I_{zd}(w-\varphi)+\cot(w+\varphi)I_{zd}(w+\varphi))+\tau_o(I_{\rho+}(w-\varphi)-I_{\rho+}(w+\varphi)))+k(V_{zd}(w-\varphi)+V_{zd}(w+\varphi))}{2k}
\end{aligned}$$

$$\begin{aligned}
&I_{zd}(w,\varphi) \\
&=\frac{kZ_o(I_{zd}(w-\varphi)+I_{zd}(w+\varphi))+\alpha_o(\cot(w-\varphi)V_{zd}(w-\varphi)-\cot(w+\varphi)V_{zd}(w+\varphi))+\tau_o(-V_{\rho+}(w-\varphi)+V_{\rho+}(w+\varphi))}{2kZ_o}
\end{aligned}$$

$$\begin{aligned}
I_{\rho+}(w,\varphi) &= \frac{1}{2kZ_o\tau_o}(kZ_o(\csc(w)\sin(\varphi)\alpha_o(-\csc(w-\varphi)I_{zd}(w-\varphi) \\
&+\csc(w+\varphi)I_{zd}(w+\varphi))+\tau_o(I_{\rho+}(w-\varphi)+I_{\rho+}(w+\varphi)))+ \\
&+(k^2+\cot(w)\cot(w-\varphi)\alpha_o^2)V_{zd}(w-\varphi)-k^2V_{zd}(w+\varphi) \\
&-\cot(w)\cot(w+\varphi)\alpha_o^2V_{zd}(w+\varphi)-\cot(w)\alpha_o\tau_oV_{\rho+}(w-\varphi)+\cot(w)\alpha_o\tau_oV_{\rho+}(w+\varphi))
\end{aligned} \tag{3.10.5}$$

These equations obtained in the w plane are very important to estimate the spectra in any direction of space, and they are cumbersome to be written in the η -plane.

3.11. Rotating wave method

While, in the impenetrable wedge scattering problem, we can solve the GWHEs by introducing one mapping (3.4.1) to obtain the classical WH formulation, in the penetrable wedge, we need multiple mappings to obtain something similar to the classical WH formulation.

In all the cases where an exact solution via the classical factorization of the penetrable wedge problem is available, this solution can be easily obtained using the w plane and by reducing the GWHE to difference equations.

From the beginning of this section, to obtain the solution of the wedge problem, we have formulated and used the representation of field components in terms of Laplace transforms defined in the η -, $\overline{\eta}$ -, w- or $\overline{w}$ -plane.

In this section, we introduce a new representation of the spectral problem in the w plane. The formulation is given by rotating waves that have similar properties to the Sommerfeld functions that are fundamental in the SM method.

Before introducing the rotating waves, we show that Helmholtz's equation in terms of $E_z(\rho,\varphi)$ ($H_z(\rho,\varphi)$) for angular regions reduces to a simple equation (Peters 1950, 1952) and (Daniele *et al.* 2017c) if we introduce the spectral quantities $\hat{V}_{zd}(w,\varphi)$ ($\hat{I}_{zd}(w,\varphi)$) defined in (3.10.3).

For the sake of simplicity, we will assume normal incidence $\alpha_o = 0$, thus $\tau_o = k$. Let us consider the wave equation in cylindrical coordinates:

$$\frac{\partial}{\partial\rho}(\rho^2\frac{\partial E_z}{\partial\rho}) - \rho\frac{\partial E_z}{\partial\rho} + \frac{\partial^2 E_z}{\partial\varphi^2} + k^2\rho^2 E_z = 0 \tag{3.11.1}$$

By applying the Laplace transform

$$V_+(s,\varphi) = \int_0^\infty E_z(\rho,\varphi)e^{-s\rho}d\rho, I_+(s,\varphi) = \int_0^\infty H_\rho(\rho,\varphi)e^{-s\rho}d\rho \tag{3.11.2}$$

along the ρ -direction with s=-jη and taking into account that ρ E_z(ρ, φ) →0 as ρ→0, we obtain

$$s(\frac{\partial^2}{\partial s^2}(sV_+(s,\varphi)-E_z(0,\varphi))+\frac{\partial}{\partial s}(sV_+(s,\varphi)-E_z(0,\varphi))+\frac{\partial^2 V_+(s,\varphi)}{\partial\varphi^2}+k^2\frac{\partial^2 V_+(s,\varphi)}{\partial s^2}=0 \tag{3.11.3}$$

Assuming null initial conditions, we have:

$$(k^2+s^2)\frac{\partial^2}{\partial s^2}V_+(s,\varphi)+3s\frac{\partial}{\partial s}V_+(s,\varphi)+V_+(s,\varphi)+\frac{\partial^2 V_+(s,\varphi)}{\partial\varphi^2}=0 \tag{3.11.4}$$

which can be alternatively written as

$$\frac{\partial}{\partial s}\left[(k^2+s^2)\frac{\partial}{\partial s}V_+(s,\varphi)\right]+\frac{\partial}{\partial s}(sV_+(s,\varphi))+\frac{\partial^2 V_+(s,\varphi)}{\partial\varphi^2}=0 \tag{3.11.5}$$

Taking into account that $\frac{\partial}{\partial s}=\frac{1}{-jk\sin w}\frac{\partial}{\partial w}$ and $k^2+s^2=k^2\sin^2 w$, (3.11.5) becomes

$$\begin{aligned}&\frac{1}{-jk\sin w}\frac{\partial}{\partial w}(k^2\sin^2 w\frac{1}{-jk\sin w}\frac{\partial}{\partial w}\hat{V}_+(w,\varphi))\\&+\frac{1}{-jk\sin w}\frac{\partial}{\partial w}(jk\cos w\hat{V}_+(w,\varphi))+\frac{\partial^2\hat{V}_+(w,\varphi)}{\partial\varphi^2}=0\end{aligned} \tag{3.11.6}$$

Introducing $\hat{V}_d(w,\varphi)=\sin w\hat{V}_+(w,\varphi)$ and after some mathematical manipulations, (3.11.5) reduces to

$$-\frac{\partial^2\hat{V}_d(w,\varphi)}{\partial^2 w}-\frac{\partial^2\hat{V}_d(w,\varphi)}{\partial^2\varphi}=0 \tag{3.11.7}$$

which has the form of the telegrapher's equation (Daniele *et al.* 2017c).

It is well known that angular transmission lines are useful to study waves in angular regions (Felsen and Marcuvitz 1973). The solution of (3.11.7) implies the presence of clockwise and counterclockwise angular rotating waves. The Sommerfeld integrals allow the introduction of only one of these two opposite rotating waves. This simplification has a counterpart: the presence in the integral representation of a complex and artificial integration path, known as the Sommerfeld contour.

To avoid the Sommerfeld contour and, at the same time, to emphasize the importance of the Laplace transforms, Daniele (2003a) introduced the rotating waves in the Laplace domain.

In order to obtain their precise definition, we first consider a two-dimensional electromagnetic field problem in terms of $E_z(\rho,\varphi)$, $H_\rho(\rho,\varphi)$ and $H_\varphi(\rho,\varphi)$ in a free-source angular region $0 \le \rho < \infty$, $\varphi_1 \le \varphi \le \varphi_2$ (Figure 3.11.1) of impedance $Z_o = 1/Y_o$.

Let us introduce the function $v(w,\varphi)$:

$$v(w,\varphi) = -k\sin w V_+(\eta,\varphi)\big|_{\eta=-k\cos(w)} = -k\hat{V}_d(w,\varphi) \tag{3.11.8}$$

From (3.11.7), $v(w,\varphi)$ satisfies the equation

$$\frac{\partial^2 v(w,\varphi)}{\partial w^2} - \frac{\partial^2 v(w,\varphi)}{\partial \varphi^2} = 0 \tag{3.11.9}$$

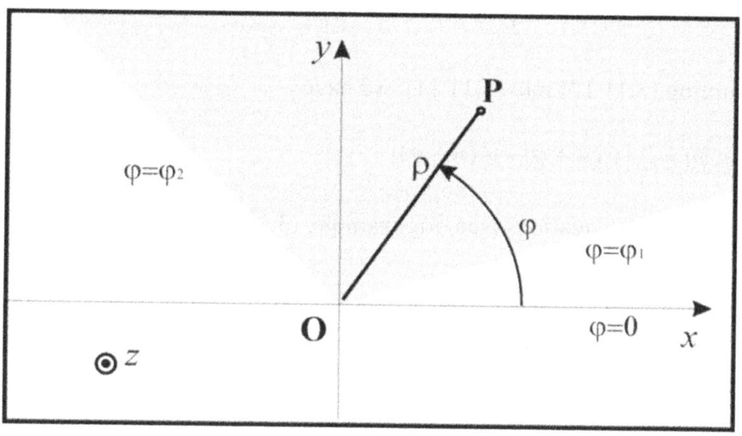

Figure 3.11.1. *Angular region* $0 \le \rho < \infty, \varphi_1 \le \varphi \le \varphi_2$

The most general solution of (3.11.9) is

$$v(w,\varphi) = v_1(w+\varphi) + v_2(w-\varphi) \tag{3.11.10}$$

where $v_1(w)$ $(v_2(w))$ is the clockwise (counterclockwise) rotating wave in the angular region.

By applying the Laplace transform (3.11.2) to the Maxwell equation $-j\rho k Z_o H_\rho = \frac{\partial E_z}{\partial \varphi}$ and introducing the w plane, we have

$$-Z_o \frac{\partial \hat{I}_+(w,\varphi)}{\partial w} = \frac{\partial \hat{V}_d(w,\varphi)}{\partial \varphi} \tag{3.11.11}$$

from the dual property of the Fourier transform of a derivative.

From (3.11.8) and (3.11.10), we obtain

$$\begin{aligned} -k\frac{\partial}{\partial \varphi}\hat{V}_d(w,\varphi) &= \frac{\partial}{\partial \varphi}v(w,\varphi) = \frac{\partial}{\partial \varphi}v_1(w+\varphi) + \frac{\partial}{\partial \varphi}v_2(w-\varphi) \\ &= \frac{\partial}{\partial w}v_1(w+\varphi) - \frac{\partial}{\partial w}v_2(w-\varphi) \end{aligned} \tag{3.11.12}$$

By defining the Laplace transform of the radial component of the magnetic field

$$i(w,\varphi) = kI_+(\eta,\varphi) = k\int_0^\infty H_\rho(\rho,\varphi)e^{j\eta\rho}\,d\rho\Big|_{\eta=-k\cos(w)} \tag{3.11.13}$$

and substituting (3.11.12) into (3.11.11), we have

$$i(w,\varphi) = Y_0\left[v_1(w+\varphi) - v_2(w-\varphi)\right] \tag{3.11.14}$$

From Maxwell's equations (see, for example, (3.11.11)) we have

$$\frac{\partial}{\partial \varphi}v(w,\varphi) = Z_o\frac{\partial}{\partial w}i(w,\varphi) \tag{3.11.15}$$

$$\frac{\partial}{\partial \varphi}i(w,\varphi) = Y_o\frac{\partial}{\partial w}v(w,\varphi) \tag{3.11.16}$$

and from (3.11.9), we can interpret the equations between $v(w,\varphi)$ and $i(w,\varphi)$ as the model of the transmission line with a phase velocity equal to 1.

It follows that the rotating waves $v_1(w+\varphi), v_2(w-\varphi)$ play the role of the forward (clockwise) and backward (counterclockwise) traveling waves of a uniform angular transmission line, where the role of the time is assumed by the complex variable w.

Let us consider a source consisting of an incident plane wave in the angular region:

$$E_z^i(\rho,\varphi) = E_o e^{jk\rho\cos(\varphi-\varphi_o)} \tag{3.11.17}$$

$$H_\rho^i(\rho,\varphi) = -\frac{1}{j\omega\mu\rho}\frac{\partial E_z^i(\rho,\varphi)}{\partial\varphi} = Y_o \sin(\varphi-\varphi_o)e^{jk\rho\cos(\varphi-\varphi_o)}E_o \tag{3.11.18}$$

In the Laplace domain, (3.11.17) and (3.11.18) become

$$V_+^i(\eta,\varphi) = \frac{j}{\eta + k\cos(\varphi-\varphi_o)}E_o \tag{3.11.19}$$

$$I_+^i(\eta,\varphi) = j\frac{Y_o\sin(\varphi-\varphi_o)}{\eta + k\cos(\varphi-\varphi_o)}E_o \tag{3.11.20}$$

We observe that the spectra (3.11.19)–(3.11.20) show the GO pole $\eta = -k\cos(\varphi = \varphi_o)$. Therefore, for an arbitrary direction of observation φ, the convergence region is at least $\text{Im}[\eta] > \text{Im}[-k]$.

In the w plane, from (3.11.8) and (3.11.13), we have for the incident wave

$$v^i(w,\varphi) = \frac{j\sin w}{\cos w - \cos(\varphi-\varphi_o)}E_o \tag{3.11.21}$$

$$i^i(w,\varphi) = \frac{jY_o\sin(\varphi-\varphi_o)}{\cos w - \cos(\varphi-\varphi_o)}E_o \tag{3.11.22}$$

Since plane wave sources originate at infinity, their presence keeps the wave equations in the homogenous forms (3.11.1) and (3.11.9), yielding solutions with just progressive and regressive waves (3.11.10) and (3.11.14).

The Laplace transforms (3.11.21) and (3.11.22) through (3.11.10) and (3.11.14), evaluated at $\varphi = 0$, yield the following rotating waves relevant to the plane wave:

$$\begin{cases} v_1^i(w) = \dfrac{j(\sin w + \sin\varphi_o)}{2(\cos w - \cos\varphi_o)}E_o \\ v_2^i(w) = \dfrac{j(\sin w - \sin\varphi_o)}{2(\cos w - \cos\varphi_o)}E_o \end{cases} \tag{3.11.23}$$

In the following, we denote the incident rotating waves by $v_1^i(w)$ and $v_2^i(w)$.

We note the validity of the property

$$v_2^i(w) = -v_1^i(-w) \tag{3.11.24}$$

3.12. Properties of rotating waves

In this section, we demonstrate several important properties of rotating waves: property a) $v_1(w) = -v_2(-w)$; property b) if $v_1(w)$ is regular in $\varphi_1 \leq \mathrm{Re}[w] \leq \varphi_2$ then $v_2(w)$ is regular in $-\varphi_2 \leq \mathrm{Re}[w] \leq -\varphi_1$; property c) $v_1(w+\varphi)$ and $v_2(w-\varphi)$ are bounded as $w \to \pm j\infty$ for any $\varphi : \varphi_1 \leq \varphi \leq \varphi_2$; property d) if the longitudinal field satisfies, near the edge $\rho = 0$, the condition $E_z(\rho,\varphi) = O(\rho^c)$ with Re[c]>0, the rotating waves behave as $v_{1,2}(w) = O[\exp(-c|\mathrm{Im}[w]|)]$ as $w \to \pm j\infty$; and property e) $v_1(w) = -j\, s(w)$, where $s(w)$ is the Sommerfeld function, if applicable.

Property a

The homogeneous region considered in Figure 3.11.1 is generally bounded by other homogeneous or non-homogeneous regions. The presence of discontinuity generates the scattered fields $E_z^s(\rho,\varphi)$ and $H_\rho^s(\rho,\varphi)$ in addition to the incident fields $E_z^i(\rho,\varphi)$ and $H_\rho^i(\rho,\varphi)$:

$$E_z^s(\rho,\varphi) = E_z(\rho,\varphi) - E_z^i(\rho,\varphi),\ H_\rho^s(\rho,\varphi) = H_\rho(\rho,\varphi) - H_\rho^i(\rho,\varphi) \tag{3.12.1}$$

Several types of waves can be present in an angular region: surface waves, leaky waves, lateral waves and so on. We recall that the Laplace transform theory assures that, in every case, the non-standard part of $V_+(\eta,\varphi)$ and $I_+(\eta,\varphi)$ is due to the GO wave poles ($\eta_{go} = -k\cos\varphi_{go}$, in particular, $\varphi_{go} = \varphi_o$) located in the segment $\eta = -ku,\ 0 \leq u \leq 1$ (corresponding to $-\pi/2 < w < 0$ for $\eta = -k\cos w$).

As commented after (3.11.20), the $V_+^i(\eta,\varphi)$ and $I_+^i(\eta,\varphi)$ can have a pole $\eta_o = -k$ in the particular case $\varphi = \varphi_o$.

Therefore, we work with the Laplace transforms of the scattered field defined by

$$V_+^s(\eta,\varphi) = L[E_z(\rho,\varphi) - E_z^i(\rho,\varphi)] = \int_0^\infty [E_z(\rho,\varphi) - E_z^i(\rho,\varphi)]e^{-s\rho}d\rho\Big|_{s=-j\eta} \tag{3.12.2}$$

$$I_+^s(\eta,\varphi) = L[H_\rho(\rho,\varphi) - H_\rho^i(\rho,\varphi)] = \int_0^\infty [H_\rho(\rho,\varphi) - H_\rho^i(\rho,\varphi)]e^{-s\rho}d\rho\Big|_{s=-j\eta} \tag{3.12.3}$$

which are always regular in the η-upper half-plane ($\mathrm{Im}[\eta] \geq \mathrm{Im}[-k]$) and, in particular, on the half-line $\eta = -ku, u \geq 1$ including the point $\eta = -k$, also in the presence of reflected waves (except for the impractical limit case $\varphi_o = 0$). On the w plane, the corresponding image of the regular line $\eta = -ku, u \geq 1$ is the imaginary positive half-axis $w = jw''$, $w'' \geq 0$.

This regularity property is fundamental to first demonstrate

$$v_2^s(w) = -v_1^s(-w) \tag{3.12.4}$$

on $w = jw"$, $w" \geq 0$ (including in $w" = 0$ that corresponds to $\eta = -k$).

This regularity property holds for all the values of φ ($\varphi_1 \leq \varphi \leq \varphi_2$), and, after some mathematical manipulations, it yields the validity of the fundamental property (3.12.4) for the scattered rotating waves for any w.

In fact, from (3.11.8) and (3.11.10), we have at w=0 ($\eta = -k$)

$$v^s(0,\varphi) = 0 = v_1^s(\varphi) + v_2^s(-\varphi) \tag{3.12.5}$$

i.e. $f(w) = v_1^s(w) + v_2^s(-w)$ is null on the segment $\varphi_1 \leq \mathrm{Re}[w] \leq \varphi_2$, $\mathrm{Im}[w] = 0$.

A process of analytic continuation (also demonstrated by series expansion) allows us to show that $f(w)$ is null in the whole complex plane w, thus (3.12.4) holds for any w.

Taking into account (3.11.24) for the incident wave and (3.12.4) for the scattered wave, the two opposite rotating waves are then related by the property:

$$v_1(w) = -v_2(-w) \tag{3.12.6}$$

Equations (3.11.8), (3.11.10) and (3.11.14), along with (3.12.6), demonstrate again, as described in section 1.14, that the plus functions are even functions of w in the w plane.

Property b

The scattered rotating waves $v_{1,2}^s(w)$ defined by

$$v_{1,2}^s(w) = v_{1,2}(w) - v_{1,2}^i(w) \tag{3.12.7}$$

are respectively regular in the strips $\mathrm{I}_1\{\varphi_1 \leq \mathrm{Re}[w] \leq \varphi_2\}$ and $\mathrm{I}_2\{-\varphi_2 \leq \mathrm{Re}[w] \leq -\varphi_1\}$ (see Figure 3.12.1).

To prove this property, let us observe that from (3.11.8)–(3.11.14), we have:

$$v_1^s(w+\varphi) = \frac{-k\sin w V_+^s(-k\cos w,\varphi) + \omega\mu I_+^s(-k\cos w,\varphi)}{2} \tag{3.12.8}$$

$$v_2^s(w-\varphi) = \frac{-k\sin w V_+^s(-k\cos w,\varphi) - \omega\mu I_+^s(-k\cos w,\varphi)}{2} \tag{3.12.9}$$

which are a particular case of (3.13.8)–(3.13.10) at normal incidence $(\beta = \pi/2)$.

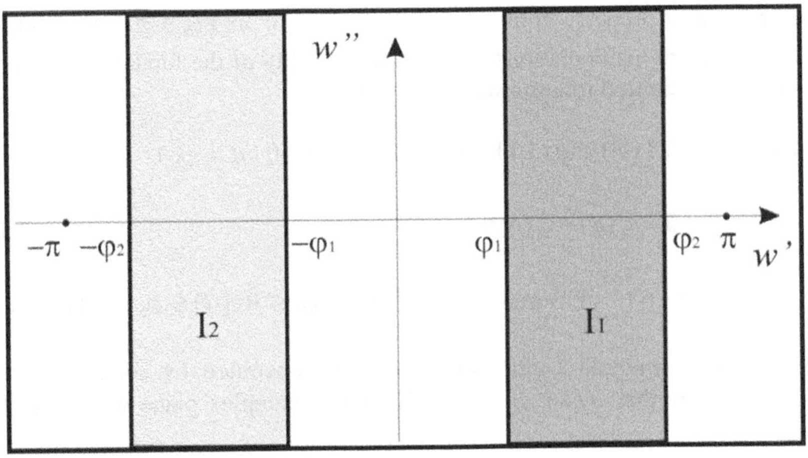

Figure 3.12.1. *Regularity regions of $v_{1,2}^s(w)$ respectively in $I_1\{\varphi_1 \leq Re[w] \leq \varphi_2\}$ and $I_2\{-\varphi_2 \leq Re[w] \leq -\varphi_1\}$*

From these equations, we ascertain that the regularity of $V_+^s(-k\cos w,\varphi)$ and $I_+^s(-k\cos w,\varphi)$ stated in property a) implies that $v_1^s(jw''+\varphi)$ and $v_2^s(jw''-\varphi)$ are also regular for all the values of $\varphi: \varphi_1 \leq \varphi \leq \varphi_2$ and w'' ≥ 0. Consequently, making

$\varphi = w'$, the analytic functions $v_1^s(w'+jw'')$ and $v_2^s(-w'+jw'')$ are holomorphic in the half-strips defined by $\varphi_1 \leq w' \leq \varphi_2$, $w'' \geq 0$.

Taking into account (3.12.6), we prove the regularity property of $v_{1,2}^s(w)$ respectively in I_1 and I_2.

Property c

The functions $v_1(w+\varphi)$ and $v_2(w-\varphi)$ are bounded as $w \to \pm j\infty$ for any $\varphi : \varphi_1 \leq \varphi \leq \varphi_2$ and the bounded values are independent from φ:

$$v_1(\pm j\infty + \varphi) = v_1(\pm j\infty), \qquad v_2(\pm j\infty + \varphi) = v_2(\pm j\infty) \tag{3.12.10}$$

To prove this property, we first observe that the initial value theorem in the w-plane may be formulated in the form

$$\lim_{\rho \to 0} f(\rho) = \lim_{\eta \to j\infty} [-j\eta F_+(\eta)] = \lim_{w \to +j\infty} \left[jk \cos w F_+(-k \cos w) \right] \tag{3.12.11}$$

Taking into account the boundedness of E_z in $\rho = 0$ and that the Laplace transform $V_+(-ku,\varphi)$ is bounded as $u \to +\infty$, it follows from (3.12.11) that $v(w,\varphi)$ is bounded as $w \to +j\infty$. At the same time, the boundedness of the Laplace transform $I_+(-ku,\varphi)$ implies that $i(w,\varphi)$ is also bounded as $w \to +j\infty$. Consequently, using (3.12.8), (3.12.9), (3.11.8) and (3.11.13), both $v_1(w+\varphi)$ and $v_2(w-\varphi)$ are bounded as $w \to +j\infty$ and assume the same values for any $\varphi : \varphi_1 \leq \varphi \leq \varphi_2$. Using property b, we extend these characteristics of $v_1(w+\varphi)$ and $v_2(w-\varphi)$ to $w \to -j\infty$.

Property d

If the longitudinal field satisfies, near the edge $\rho = 0$, the condition $E_z(\rho,\varphi) = O(\rho^c)$ with Re[c]>0, the rotating waves behave as

$$v_{1,2}(w) = O[\exp(-c|\mathrm{Im}[w]|)] \text{ as } w \to \pm j\infty \tag{3.12.12}$$

This property also follows from the behavior of the Laplace transforms at $w \to \pm j\infty$.

Using property a, it is sufficient to study this behavior for $w \to j\infty$.

If $E_z = O(\rho^{c})$ and $H_\rho = O(\rho^{c-1})$, it follows from Watson's lemma (Jones 1964 p. 438) that

$$V_+(\eta,\varphi) = O(\eta^{-(c+1)}) \text{ and } I_+(\eta,\varphi) = O(\eta^{-c}) \tag{3.12.13}$$

thus

$$\sin w V_+(-k\cos w,\varphi) = O\left(e^{|\mathrm{Im}[w]|}\right) O\left(e^{-|\mathrm{Im}[w]|(c+1)}\right) = O\left(e^{-|\mathrm{Im}[w]|c}\right) \tag{3.12.14}$$

$$I_+(-k\cos w,\varphi) = O\left(e^{-|\mathrm{Im}[w]|c}\right) \tag{3.12.15}$$

Substituting (3.12.14) and (3.12.15) into (3.12.8) and (3.12.9) completes the proof.

Property e

When the Sommerfeld integral

$$E_z(\rho,\varphi) = \frac{1}{2\pi j}\left[\int_\gamma s[w+\varphi] e^{+jk\cos[w]\rho} dw\right] \tag{3.12.16}$$

is valid, the Sommerfeld function $s(w)$ is related to the clockwise wave $v_1(w)$ through the equation

$$v_1(w) = -j\, s(w) \tag{3.12.17}$$

We prove (3.12.17) by comparing the inverse Laplace integral

$$E_z(\rho,\varphi) = \frac{1}{2\pi}\int_\gamma V_+(\eta,\varphi) e^{-j\eta\rho} d\eta = -\frac{1}{2\pi}\int_{B_r} v(w,\varphi) e^{jk\rho\cos w} dw \tag{3.12.18}$$

$$E_z(\rho,\varphi) = -\frac{1}{2\pi}\int_{B_r} [v_1(w+\varphi) + v_2(w-\varphi)] e^{jk\rho\cos w} dw \tag{3.12.19}$$

with the Sommerfeld integral (Senior and Volakis 1995)

$$E_z(\rho,\varphi) = \frac{1}{2j\pi}\left[\int_{C_1\cup C_2} s[w+\varphi] e^{+jk\cos w\rho} dw\right] \tag{3.12.20}$$

In (3.12.18), γ is the Bromwich contour in the η-plane and B_r is its image in the w plane.

In (3.12.20), $C = C_1 \bigcup C_2$ is the Sommerfeld contour (see Figure 3.12.2, and Chapter 4 of (Daniele and Zich 2014)).

Due to the regularity of $V_+(\eta,\varphi)$ in $\text{Im}[\eta] \geq \text{Im}[-k]$, we can assume γ as a horizontal line cutting the branch line b_- relevant to the branch point $-k$.

$V_+(\eta,\varphi)$ is regular on this branch line, hence B_r is continuous, as shown in Figure 3.12.2. Therefore, we can warp B_r into C_2. Finally, taking into account that the substitution $w \to -w$ transforms C_2 into C_1, from (3.12.19) and considering (3.12.6), we obtain

$$E_z(\rho,\varphi) = -\frac{1}{2\pi}\int_{C_2}[v_1(w+\varphi) - v_1(\varphi - w)]e^{jk\rho\cos w}dw = -\frac{1}{2\pi}\int_{C_1 \cup C_2} v_1(w+\varphi)e^{jk\rho\cos w}dw \tag{3.12.21}$$

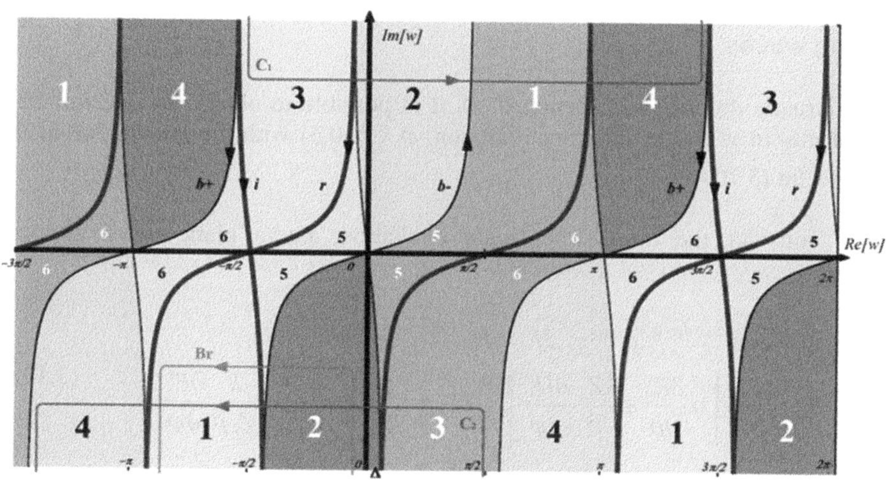

Figure 3.12.2. *Images of the proper and improper sheets of the η-plane in the w plane are respectively shown with gray and blue colors. For the proper sheet, the images of the four quadrants 1, 2, 3 and 4 are delimited by the images of the standard branch lines b+ and b-. The r and i lines are the images of the real and imaginary axes of the η-plane. Sub-regions 5 and 6 are respectively part of the second and fourth quadrants of the η-plane, but they show a different sign on the real part of ξ with respect to the rest of their quadrants. In this figure, k is equal to 1-j0.2. Δ is $\arctan(-Im[k]/Re[k])$. The proper and improper sheets of the η-plane are replicated periodically in the w plane with period 2π. The Bromwich contour B_r is indicated in green for the inverse Laplace transform (3.12.18) and the Sommerfeld contours C_1 and C_2 in red for the integral (3.12.20). For a color version of the figure, see www.iste.co.uk/daniele/scattering1.zip*

By comparing (3.12.20) with (3.12.21), it yields (3.12.17).

However, we cannot exclude *a priori* the presence of infinite poles between the contours B_r and C_2. In this case, even assuming a different Bromwich contour with respect to the one presented in Figure 3.12.2, we cannot immediately warp B_r into C_2.

We conclude that, in the presence of a far source consisting of a plane wave with the incoming direction φ_o, we associate incident rotating waves, as described in (3.11.23). The incident wave generates the pole $w = \varphi_o$ in the regularity strip I_1 of the clockwise wave $v_1(w)$ and the pole $w = -\varphi_o$ in the regularity strip I_2 of the counterclockwise wave $v_2(w)$. However, the properties a and c hold again, when considering these rotating wave contributions.

3.13. Spectral field component in w for an arbitrary direction φ using rotating waves

As already described in section 3.10, it is possible to obtain the spectral field components in w for an arbitrary direction φ (3.10.3) with the knowledge of the axial spectra (3.10.4).

We consider the three-dimensional excitation, i.e. a plane wave at skew incidence:

$$E_z^i = E_o e^{j\tau\rho\cos(\varphi-\varphi_o)} e^{-j\alpha z}, \quad H_z^i = H_o e^{j\tau\rho\cos(\varphi-\varphi_o)} e^{-j\alpha z}$$
$$E_\rho^i = \frac{1}{j\tau^2}\left[\alpha\frac{\partial E_z^i}{\partial\rho} + \frac{kZ_o}{\rho}\frac{\partial H_z^i}{\partial\varphi}\right], \quad H_\rho^i = \frac{1}{j\tau^2}\left[\alpha\frac{\partial H_z^i}{\partial\rho} - \frac{k}{Z_o\rho}\frac{\partial E_z^i}{\partial\varphi}\right] \qquad (3.13.1)$$

where $\alpha = k\cos\beta$ and $\tau = k\sin\beta$ with the skewness zenithal angle β and the azimuthal angle φ_o.

Consequently, in general, both the longitudinal components E_z and H_z are present in the total field.

Again, we introduce the rotating waves for the Laplace transform of the longitudinal components:

$$E_z \Rightarrow v(w,\varphi) = -\tau\sin w V_{z+}(-\tau\cos w,\varphi) = v_1(w+\varphi) + v_2(w-\varphi) \qquad (3.13.2)$$

$$H_z \Rightarrow i(w,\varphi) = -\tau \sin w I_{z+}(-\tau \cos w, \varphi) = i_1(w+\varphi) + i_2(w-\varphi) \tag{3.13.3}$$

where we have introduced the Laplace transforms $V_{z+}(\eta,\varphi)$ and $I_{z+}(\eta,\varphi)$, respectively, for the longitudinal electric and magnetic fields, and omitted the $e^{-j\alpha z}$ dependence.

Considering the relations between the radial and longitudinal components of electromagnetic fields:

$$\rho E_\rho = \frac{k}{j\tau^2}\left[\frac{\alpha}{k}\rho\frac{\partial E_z}{\partial \rho} + Z_o \frac{\partial H_z}{\partial \varphi}\right] \tag{3.13.4}$$

$$\rho H_\rho = \frac{k}{j\tau^2}\left[\frac{\alpha}{k}\rho\frac{\partial H_z}{\partial \rho} - Y_o \frac{\partial E_z}{\partial \varphi}\right] \tag{3.13.5}$$

we deduce the following Laplace transform of the radial components in terms of (3.13.2) and (3.13.3). It yields

$$\begin{aligned}\hat{V}_{\rho+}(w,\varphi) &= L\left[E_\rho\right]_{\eta=-\tau\cos w} \\ &= \left[-\frac{\omega\mu}{\tau^2}\left(i_1(w+\varphi) - i_2(w-\varphi)\right) - \frac{\alpha}{\tau^2 \tan w}\left(v_1(w+\varphi) + v_2(w-\varphi)\right)\right]\end{aligned} \tag{3.13.6}$$

$$\begin{aligned}\hat{I}_{\rho+}(w,\varphi) &= L\left[H_\rho\right]_{\eta=-\tau\cos w} \\ &= \left[\frac{\omega\varepsilon}{\tau^2}\left(v_1(w+\varphi) - v_2(w-\varphi) - \frac{\alpha}{\tau^2 \tan w}\left(i_1(w+\varphi) + i_2(w-\varphi)\right)\right)\right]\end{aligned} \tag{3.13.7}$$

by omitting the $e^{-j\alpha z}$ dependence and exploiting the property to represent the partial derivative in φ in terms of the partial derivative in w, as already done in (3.11.11)–(3.11.14).

Consider an angular homogeneous region $\varphi_1 \le \varphi \le \varphi_2$ (Figure 3.13.1) that comprises the direction $\varphi = 0$.

By putting $\varphi = 0$ in (3.13.2), (3.13.3), (3.13.6) and (3.13.7), we obtain the rotating waves $v_1(w), v_2(w), i_1(w), i_2(w)$

$$v_1(w) = -\frac{1}{2}k_o \sin\beta(Z_o \cos w \cos\beta \hat{I}_{z+}(w) - Z_o \sin\beta \hat{I}_{\rho+}(w) + \sin(w)\hat{V}_{z+}(w)) \tag{3.13.8}$$

$$i_1(w) = -\frac{k_o \sin\beta(Z_o \sin w\, \hat{I}_{z+}(w) - \cos\beta\cos w \hat{V}_{z+}(w) + \sin(\beta)\hat{V}_{\rho+}(w))}{2Z_o} \quad (3.13.9)$$

in terms of the axial spectra (where the second variable in the spectra is omitted).

Exploiting the property of rotating waves

$$v_2(w) = -v_1(-w)\,,\; i_2(w) = -i_1(-w) \quad (3.13.10)$$

from (3.13.2), (3.13.3), (3.13.6) and (3.13.7), we directly obtain the spectra for any $\varphi \neq 0$, $\varphi_1 \leq \varphi \leq \varphi_2$ in terms of shifted rotating waves and shifted axial spectra, as in section 3.10.

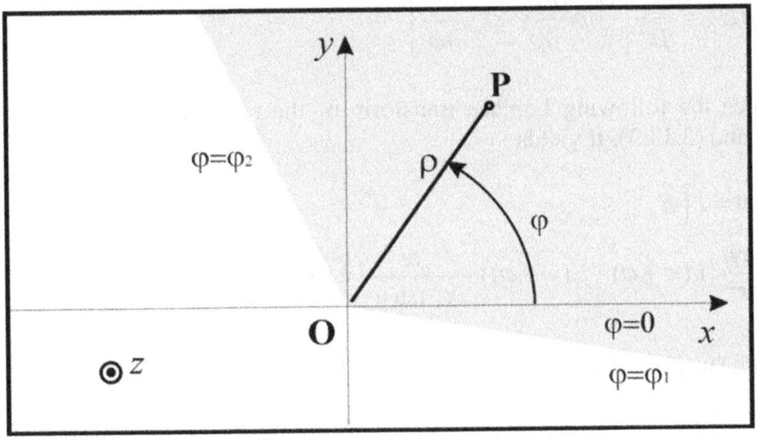

Figure 3.13.1. *Angular region* $0 \leq \rho < \infty$, $\varphi_1 \leq \varphi \leq \varphi_2$ *and negative* φ_1

3.14. Rotating waves in areas different from electromagnetism

The demonstrated benefit of rotating waves for the study of electromagnetic scattering in angular regions can be extended to wave motion problems in other physics.

In this section, we extend the definition of rotating waves for an arbitrary wave field described by a scalar quantity $u(\rho,\varphi)$ and a vector quantity $\mathbf{T}(\rho,\varphi)$ with

$$\mathbf{T} = q\nabla_t u \quad (3.14.1)$$

where q is a parameter of the medium on which the wave motions are generated.

For the sake of simplicity, we only consider two-dimensional cases (skewness angle $\beta = \pi / 2$, thus $\alpha_o = 0$).

Let us assume that $u(\rho, \varphi)$ satisfies the Helmholtz equation:

$$\nabla_t^2 u(\rho, \varphi) + k^2 u(\rho, \varphi) = 0 \tag{3.14.2}$$

In cylindrical coordinates, we have

$$\frac{\partial}{\partial \rho}(\rho^2 \frac{\partial u}{\partial \rho}) - \rho \frac{\partial u}{\partial \rho} + \frac{\partial^2 u}{\partial \varphi^2} + k^2 \rho^2 u = 0 \tag{3.14.3}$$

By decomposing $\mathbf{T}(\rho, \varphi) = \hat{\rho} T_\rho(\rho, \varphi) + \hat{\varphi} T_\varphi(\rho, \varphi)$, we have, from (3.14.1)

$$T_\rho(\rho, \varphi) = q \frac{\partial u(\rho, \varphi)}{\partial \rho} \tag{3.14.4}$$

$$T_\varphi(\rho, \varphi) = q \frac{\partial u(\rho, \varphi)}{\rho \partial \varphi} \tag{3.14.5}$$

Introducing the Laplace transforms

$$I_{\varphi +}(\eta, \varphi) = \int_0^\infty u(\rho, \varphi) e^{j\eta\rho} d\rho \tag{3.14.6}$$

in the w plane (defined by $\eta = -k \cos w$), we obtain, from (3.14.3)

$$\frac{\partial^2 \hat{I}_d(w, \varphi)}{\partial w^2} - \frac{\partial^2 \hat{I}_d(w, \varphi)}{\partial \varphi^2} = 0 \tag{3.14.7}$$

where:

$$\hat{I}_d(w, \varphi) = \sin w I_{\varphi +}(-k \cos w, \varphi) \tag{3.14.8}$$

It follows the definitions of the rotating waves $i_1(w)$ and $i_2(w)$:

$$\hat{I}_d(w, \varphi) = i_1(w + \varphi) + i_2(w - \varphi) \tag{3.14.9}$$

Defining

$$V_{\varphi+}(\eta,\varphi) = \int_0^\infty T_\varphi(\rho,\varphi)e^{j\eta\rho}d\rho \tag{3.14.10}$$

into (3.14.5), we obtain

$$\frac{\partial}{\partial w}\hat{V}_+(w,\varphi) = jkq\frac{d}{d\varphi}\hat{I}_d(w,\varphi) = jkq(\frac{\partial}{\partial w}[i_1(w+\varphi)-i_2(w-\varphi)] \tag{3.14.11}$$

i.e.

$$\hat{V}_+(w,\varphi) = jkq[i_1(w+\varphi)-i_2(w-\varphi)] \tag{3.14.12}$$

Therefore, the general wave motion problem for an angular region, where the wave fields are described by $u(\rho,\varphi)$ and $\mathbf{T}(\rho,\varphi)$, has the same representation of the electromagnetic scattering problem in terms of rotating waves.

In electromagnetics, the fundamental equations that allow the introduction of the WH technique in wedge problems with an arbitrary angle are equations (3.3.57)–(3.3.64). In this chapter, these equations are obtained using oblique coordinates. Alternative techniques to obtain the same equations are reported in (Daniele 2004), applying the Green function technique together with the properties of the Laplace transform and so on.

Moreover, the same equations can be obtained by introducing rotating waves, as reported in (Daniele 2001).

Therefore, equations (3.3.57)–(3.3.64) are valid for the general wave motion problem where the wave fields are described by $u(\rho,\varphi)$ and $\mathbf{T}(\rho,\varphi)$.

3.15. Closed-form solution of the diffraction of an elastic SH wave by wedge with classical factorization

The most simple diffraction problems in the elasticity are those with anti-planar shear motion (see sections 2.9–2.12 for an introduction on the topic). These problems are two-dimensional and the only non-vanishing component of the velocity is $v_3 = v_z$.

As a simple application of the property described in the previous section, here we study the diffraction of an SH wave by an elastic solid wedge terminated by free space (Figure 3.15.1).

In this problem, we consider the Cartesian and cylindrical coordinates, with z as the symmetry axis. We assume that the aperture angle of the elastic solid is 2Φ.

Since we have only a non-null component $v_3 = v_z$ for the velocity, the stress tensor **T** only presents the non-vanishing components $T_{z\rho}$ and $T_{z\varphi}$ defined by

$$T_{z\rho}(\rho,\varphi) = \frac{\mu}{j\omega}\frac{\partial v_3(\rho,\varphi)}{\partial\rho} \tag{3.15.1}$$

$$T_{z\varphi}(\rho,\varphi) = \frac{\mu}{j\omega}\frac{\partial v_3(\rho,\varphi)}{\rho\partial\varphi} \tag{3.15.2}$$

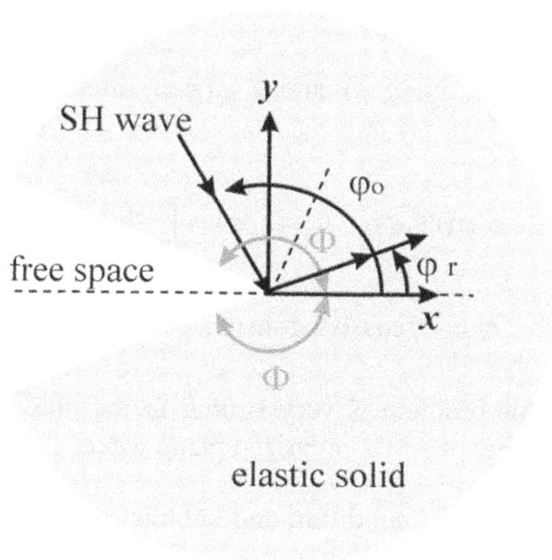

Figure 3.15.1. *Elastic solid wedge illuminated by an SH wave. For a color version of the figure, see www.iste.co.uk/daniele/scattering1.zip*

Indicating with k_{SH} the SH propagation constant of the elastic medium, we need to solve the wave equation

$$\nabla_t^2 v_3(\rho,\varphi) + k_{SH}^2 v_3(\rho,\varphi) = 0 \tag{3.15.3}$$

with the boundary conditions

$$\hat{\varphi}\cdot\mathbf{T}_z(\rho,\pm\Phi) = 0 \text{ or } T_{z\varphi}(\rho,\pm\Phi) = 0 \tag{3.15.4}$$

The possibility of introducing rotating waves for these fields (section 3.14) allows us to extend equations (3.3.58) and (3.3.60) to this case (section 3.3), which yields

Region 1 ($0 \le \varphi \le \Phi$)

$$\xi\, \hat{v}_{3+}(\eta,0) + \frac{\omega}{\mu}\hat{T}_{z\varphi+}(\eta,0) = -n\, \hat{v}_{3+}(-m,\Phi) + \frac{\omega}{\mu}\hat{T}_{z\varphi+}(-m,\Phi) \tag{3.15.5}$$

Region 2 ($-\Phi \le \varphi \le 0$)

$$\xi\, \hat{v}_{3+}(\eta,0) - \frac{\omega}{\mu}\hat{T}_{z\varphi+}(\eta,0) = -n\, \hat{v}_{3+}(-m,-\Phi) - \frac{\omega}{\mu}\hat{T}_{z\varphi+}(-m,-\Phi) \tag{3.15.6}$$

In the above equations, $\hat{v}_{3+}(\eta,\varphi)$ and $\hat{T}_{z\varphi+}(\eta,\varphi)$ are, respectively, the Laplace transform

$$\hat{v}_{3+}(\eta,\varphi) = \int_0^\infty v_3(\rho,\varphi)e^{j\eta\rho}d\rho\,,\ \hat{T}_{z\varphi+}(\eta,\varphi) = \int_0^\infty T_{z\varphi}(\rho,\varphi)e^{j\eta\rho}d\rho \tag{3.15.7}$$

and $\xi = \sqrt{k_{SH}^2 - \eta^2}$, $m = -\eta\cos\Phi + \xi\sin\Phi$, $n = -\eta\sin\Phi - \xi\cos\Phi$.

We observe that this problem is very similar to the one solved in section 3.8: scattering by a PEC wedge of an Ez-polarized plane wave.

By imposing the boundary condition and subtracting the two GWHEs (3.15.5) and (3.15.6), we obtain

$$2\frac{\omega}{\mu}\hat{T}_{z\varphi+}(\eta,0) = n[\hat{v}_{3+}(-m,-\Phi) - \hat{v}_{3+}(-m,+\Phi)] = nA_+(-m) \tag{3.15.8}$$

By applying the mapping (3.4.1), the GWHE (3.15.8) is reduced to a CWHE with the scalar kernel

$$2\frac{\omega}{\mu\bar{n}(\bar{\eta})}\bar{X}_+(\bar{\eta}) = \bar{X}_-(\bar{\eta}) \tag{3.15.9}$$

where $\bar{X}_+(\bar{\eta}) = \bar{\hat{T}}_{z\varphi+}(\bar{\eta},0)$, $\bar{X}_-(\bar{\eta}) = A_+(-\bar{m}(\bar{\eta}))$.

The factorization of $\bar{n}(\bar{\eta}) = n(\eta(\bar{\eta}))$ can be obtained for inspection as done for the PEC wedge at normal incidence in section 3.8, or using the general method indicated in the Appendix of Chapter 4. We obtain

$$\bar{n}_+(\bar{\eta}) = \sqrt{\frac{k_{SH}^2 - \bar{\eta}}{2}}, \ \bar{n}_-(\bar{\eta}) = \bar{n}\,(\bar{\eta}) / \bar{n}_+(\bar{\eta}) \tag{3.15.10}$$

Once the GO component of the plus function $\bar{T}_o$ is extracted (as done for the PEC wedge in section 3.8), the solution is thus obtained from (3.15.9) and (3.15.10) as

$$\bar{X}_+(\bar{\eta}) = \frac{\bar{n}_+(\bar{\eta})[\bar{n}_+(\bar{\eta}_o)]^{-1}\bar{T}_o}{\bar{\eta} - \bar{\eta}_o} \tag{3.15.11}$$

Using the general procedure indicated in sections 2.9–2.12, we obtain the elastic fields at any point of space.

3.16. Rotating waves with the MF transform for wedge problems

The formulation of wedge problems in the w plane yields difference equations.

When possible, the solution of difference equations is performed using the MF transforms (see section 3.9 for a preliminary use and Appendix 3.A for the definition):

$$X(t) = -j\int_{-j\infty}^{j\infty} x(w)\exp[jt\,w]dw, \ \mathrm{Re}[t] = 0 \tag{3.16.1}$$

$$x(w) = -\frac{j}{2\pi}\int_{-j\infty}^{j\infty} X(t)\exp[-jt\,w]dt, \ \mathrm{Re}[w] = 0 \tag{3.16.2}$$

For Re[t]=0, $\varphi_1 \le \varphi \le \varphi_2$, we have the following transforms of the rotating waves:

$$V_{1,2}[t,\varphi] = -j\int_{-j\infty}^{j\infty}\left[v_{1,2}(w \pm \varphi)\right]\exp[+jt\,w]dw \tag{3.16.3}$$

and the total fields:

$$V(t,\varphi) = V_1(t,\varphi) + V_2(t,\varphi), \ I(t,\varphi) = Y_0[V_1(t,\varphi) - V_2(t,\varphi)] \tag{3.16.4}$$

It should be observed that, in general, the rotating waves may be bounded but not vanishing for $w \to \pm j\infty$. Consequently, we must define Fourier transforms and inverse MF transforms in the distribution space (Appendix 3.A).

After some mathematical manipulations, applying (3.16.1) to the incident rotating waves (3.11.23) yields

$$V_1^i(t,\varphi) = -\pi \frac{e^{jt(\varphi_o - \pi - \varphi)}}{\sin(\pi t)} E_o - 2j\pi E_o\, e^{jt(-\varphi+\varphi_o)}[u(-\varphi_o) + u(\varphi - \varphi_o)u(\varphi_o)], \quad \varphi > 0 \tag{3.16.5}$$

$$V_1^i(t,\varphi) = -\pi \frac{e^{jt(\varphi_o - \pi - \varphi)}}{\sin(\pi t)} E_o - 2j\pi E_o\, e^{jt(-\varphi+\varphi_o)}[u(-\varphi_o) - u(-\varphi + \varphi_o)u(-\varphi_o)], \quad \varphi < 0 \tag{3.16.6}$$

with

$$V_2^i(t,\varphi) = -V_1^i(-t,\varphi) \tag{3.16.7}$$

The regularity of the scattered waves in the strips I_1 and I_2 allows the use of the following transport theorem (Appendix 3.A):

$$V_{1,2}^s(t,\varphi_b) = \exp[\mp jt(\varphi_b - \varphi_a)] V_{1,2}^s(t,\varphi_a) \tag{3.16.8}$$

where $\varphi_1 \le \varphi_{a,b} \le \varphi_2$.

This theorem yields the transformation of the derivative in w into an algebraic multiplication in t $\partial/\partial w \Rightarrow -jt$ and, consequently, the transmission line equations (3.11.15) and (3.11.16) in the t domain become (Daniele *et al.* 2017c)

$$-\frac{\partial}{\partial\varphi} V^s(t,\varphi) = jt\, Z_o\, I^s(v,\varphi) \tag{3.16.9}$$

$$-\frac{\partial}{\partial\varphi} I^s(t,\varphi) = jt\, Y_o\, V^s(v,\varphi) \tag{3.16.10}$$

3.17. Alternative solution of PEC wedge scattering problems via difference equations and the MF transform in terms of rotating waves

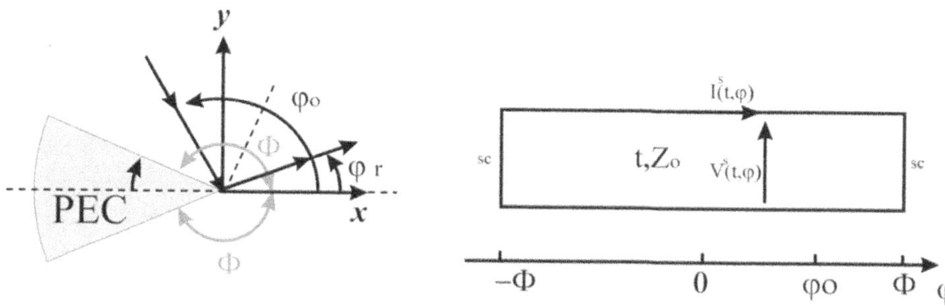

Figure 3.17.1. *Left: PEC wedge illuminated by a plane wave at Ez polarization. Right: transmission line model in the t domain for the scattered wave. For a color version of the figure, see www.iste.co.uk/daniele/scattering1.zip*

In sections 3.8 and 3.9, we solved the PEC wedge at normal incidence using the WH technique and the difference equations that result from the WH equations in the w plane.

In this section, we propose the solution of this problem using the rotating waves.

In Figure 3.17.1, the PEC wedge is immersed in a homogeneous isotropic medium defined by the angular region $-\Phi \leq \varphi \leq \Phi$. The source is an Ez-polarized plane wave propagating in the direction φ_o (3.11.17). Taking into account the PEC boundary conditions $V(t,\pm\Phi)=0$ and using the transmission line theory from (3.16.9) and (3.16.10), we obtain

$$\begin{vmatrix} e^{-jt\Phi} & e^{jt\Phi} \\ e^{jt\Phi} & e^{-jt\Phi} \end{vmatrix} \begin{vmatrix} V_1^s(t,0) \\ V_2^s(t,0) \end{vmatrix} = -\begin{vmatrix} V^i(t,\Phi) \\ V^i(t,-\Phi) \end{vmatrix} = -\begin{vmatrix} V_1^i(t,\Phi)+V_1^i(t,\Phi) \\ V_2^i(t,\Phi)+V_2^i(t,\Phi) \end{vmatrix} \tag{3.17.1}$$

since the total scattered (progressive and regressive waves) need to compensate the incident at both $\varphi=\pm\Phi$. Substituting this into the third member (3.16.5)–(3.16.7), it yields

$$V_1^s(t,0) = \pi \frac{e^{jt(\varphi_o-\pi)}}{\sin(\pi t)} E_o + 2\pi j E_o \frac{e^{-j\Phi t}\sin((\Phi-\varphi_o)t)}{\sin(2\Phi t)} \tag{3.17.2}$$

$$V_2^s(t,0) = -V_1^s(-t,0) \tag{3.17.3}$$

By applying the inverse MF transform (3.16.2), we have (Daniele 2003a)

$$v_1(w) = v_1^i(w) + v_1^s(w) = -\frac{j}{n}\frac{\cos\frac{\varphi_o}{n}}{\sin\frac{w}{n} - \sin\frac{\varphi_o}{n}} E_o = -js(w) \tag{3.17.4}$$

where $n = \frac{2\Phi}{\pi}$ and $s(w)$ is the Sommerfeld function.

Taking into account that the counterclockwise wave is $v_1(w) = -v_2(-w)$ with (3.17.4), by applying (3.12.16)–(3.12.18), we obtain the well-known result

$$\begin{aligned} E_z(\rho,\varphi) &= \frac{1}{2\pi}\int_\gamma V_+(\eta,\varphi)e^{-j\eta\rho}d\eta \\ &= -\frac{1}{2\pi}\int_{Br} v(w,\varphi)e^{jk\rho\cos w}dw = \frac{1}{2j\pi}\left[\int_{C_1\cup C_2} s[w+\varphi]e^{+jkCos[w]\rho}dw\right] \end{aligned} \tag{3.17.5}$$

$$E_z(\rho,\varphi) = \frac{jE_o}{2\pi n}\cos\frac{\varphi_o}{n}\int_{B_r}\left(\frac{1}{\sin\frac{w+\varphi}{n} - \sin\frac{\varphi_o}{n}} + \frac{1}{\sin\frac{w-\varphi}{n} + \sin\frac{\varphi_o}{n}}\right)e^{jk\rho\cos w}dw \tag{3.17.6}$$

where γ is the Bromwich contour, B_r is its image in the complex plane w through the mapping $\eta = -k\cos w$ and $C_1 \cup C_2$ is the Sommerfeld contour (see Figure 3.12.2).

Appendix 3.A: the Malyuzhinets–Fourier (MF) transform

In order to exploit the solutions in the w plane with the Laplace transforms $\hat{F}(w)$, it is useful to introduce the MF transform defined by:

$$M\{\hat{F}(w)\} = \bar{F}[t] = -j\int_{-j\infty}^{j\infty}\left[\hat{F}(w)\right]\exp[+jtw]dw, \quad \text{Im}[t] = 0 \tag{3.A.1}$$

$$\hat{F}[w] = M^{-1}\{\bar{F}(t)\} = -\frac{j}{2\pi}\int_{-j\infty}^{j\infty}\left[\bar{F}(t)\right]\exp[-jtw]dt, \quad \text{Im}[w] = 0 \tag{3.A.2}$$

This transform must be defined in the distribution space; thus, the integrals in (3.A.1) and (3.A.2) are Cauchy principal values.

In particular, we note that while the Laplace transforms are vanishing at infinity, the MF transforms goes to a constant at infinity.

Relations between the natural domain in ρ, the Laplace transform (LT) in η, the complex plane w ($\eta = -k\cos w$), the Malyuzhinets–Fourier (MF) transform in t, the Kontorovich–Lebedev (KL) transform and the Sommerfeld–Malyuzhinets (SM) function have been studied in Malyuzhinets (1958b) and Daniele (2003a).

We graphically summarize these relations in Figure 3.A.1.

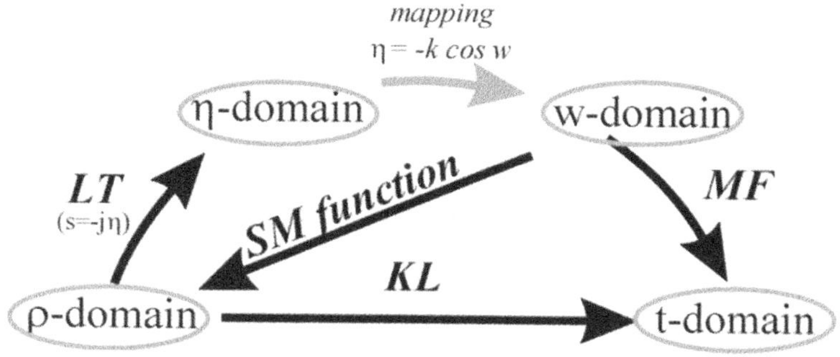

Figure 3.A.1. *Graphical representation of natural and complex spectral planes with related transforms (LT, SM, KL, MF) and the mapping* $\eta = -k\cos w$

By applying the MF transform to Maxwell equations (3.11.11) (Faraday)

$$-Z_o \frac{\partial \hat{I}_+(w,\varphi)}{\partial w} = \frac{\partial \hat{V}_d(w,\varphi)}{\partial \varphi} \tag{3.A.3}$$

and the analogous (Ampere)

$$-Y_o \frac{\partial \hat{V}_d(w,\varphi)}{\partial w} = \frac{\partial \hat{I}_+(w,\varphi)}{\partial \varphi} \tag{3.A.4}$$

it yields

$$\frac{\partial \overline{V}_d(t,\varphi)}{\partial \varphi} = jtZ_o\overline{I}_+(t,\varphi) \tag{3.A.5}$$

$$\frac{\partial \overline{I}_+(t,\varphi)}{\partial \varphi} = jtY_o\overline{V}_d(t,\varphi) \tag{3.A.6}$$

Equations (3.A.5) and (3.A.6) are the classical equations of transmission lines in the t-plane along the azimuthal direction φ. The possibility of resorting to transmission lines is very useful for producing network equivalent representations of the angular regions (Daniele *et al.* 2017c).

In this book, we limit the applications of the MF transforms to the solution of the difference equations that arise in wedge problems. Some examples of these applications were considered in the previous sections.

In this context, the application of rotating waves is more appropriate than the use of Laplace transforms of field components in the w plane.

For Re[t]=0, we define the following MF transforms of the rotating waves $(\varphi_1 \le \varphi \le \varphi_2)$:

$$V_1(t,\varphi) = -j\int_{-j\infty}^{j\infty} v_1(w+\varphi)e^{jtw}dw, \qquad \varphi_1 \le \text{Re}[w] \le \varphi_2 \tag{3.A.7}$$

$$V_2(t,\varphi) = -j\int_{-j\infty}^{j\infty} v_2(w-\varphi)e^{jtw}dw, \qquad -\varphi_2 \le \text{Re}[w] \le -\varphi_1 \tag{3.A.8}$$

and the inverse transforms

$$v_1(w+\varphi) = -\frac{j}{2\pi}\int_{-j\infty}^{j\infty} V_1(t,\varphi)e^{-jtw}dt, \qquad \varphi_1 \le \text{Re}[w] \le \varphi_2 \tag{3.A.9}$$

$$v_2(w-\varphi) = -\frac{j}{2\pi}\int_{-j\infty}^{j\infty} V_2(t,\varphi)e^{-jtw}dt, \qquad -\varphi_2 \le \text{Re}[w] \le -\varphi_1 \tag{3.A.10}$$

Again, the above integrals involve distributions spaces. In particular, we note that, while the Laplace transforms are always vanishing in the vertical strips of the w plane, the rotating waves are bounded but can be non-vanishing (see, for example, (3.11.23) for the incident plane wave). We recall that (3.A.11) holds for the values of φ belonging to the regularity strips I_1 and I_2 (see section 3.12):

$$v_1(\pm j\infty+\phi) = v_1(\pm j\infty), \qquad v_2(\pm j\infty+\phi) = v_2(\pm j\infty) \tag{3.A.11}$$

Taking into account that Laplace transforms are always vanishing, from (3.11.13)–(3.11.14) for $w = j\infty, \varphi = 0$, we obtain the following property for rotating waves:

$$i(\pm j\infty,0) = Y_0\left[v_1(\pm j\infty) - v_2(\pm j\infty)\right] = 0 \tag{3.A.12}$$

thus

$$v_{1,2}(j\infty)+v_{1,2}(-j\infty)=0 \tag{3.A.13}$$

These properties make the rotating waves particularly useful for the solution of the difference equations derived from the transport theorem:

$$V_1(t,\varphi_b)=\mathrm{e}^{-jt(\varphi_b-\varphi_a)}\,V_1(t,\varphi_a), \qquad \varphi_1\leq\varphi_a,\varphi_b\leq\varphi_2 \tag{3.A.14}$$

$$V_2(t,\varphi_b)=\mathrm{e}^{+jt(\varphi_b-\varphi_a)}\,V_2(t,\varphi_a), \qquad -\varphi_2\leq\varphi_a,\varphi_b\leq-\varphi_1 \tag{3.A.15}$$

Note that the transport theorem is valid in MF transforms, although they are not vanishing at infinity, as in Laplace transforms, that are vanishing at infinity.

We recall that these properties hold in the regularity regions although the rotating waves are not vanishing at infinity.

In general, the introduction of the MF transform in difference equations of the w-plane with constant coefficients yields an algebraic system in the t-domain:

$$\mathrm{P(t)\cdot X(t)=N_o(t)} \tag{3.A.16}$$

The unknown vector X(t) is a tempered distribution defined in Re[t]=0.

In general, the solution of the previous system yields, in the distribution space,

$$X(t)=P^{-1}(t)N_o(t)+\sum_m K_m\delta(jt-z_m) \tag{3.A.17}$$

where z_m are the simple zeros of Det[P(t)] that are possibly located on the imaginary axis Re[t]=0 (to clarify, we make a simple academic example: $xf(x)=0$ has the solution $f(x)=c\delta(x)$ in the distribution space).

X(t) must vanish for $t\to\pm j\infty$, otherwise there will be impulsive functions in the w-domain.

In addition, we can only have $z_m=0$, otherwise the inverse transform of $\delta(jt-z_m)$ will give functions $e^{-j(w\pm\phi)z_m}$ that are not bounded for $w\to\pm j\infty$.

Consequently, the solution in the t-plane always assumes the form

$$X(t) = \frac{R_o}{t} + \left\{ P^{-1}(t)N_o(t) - \frac{R_o}{t} \right\} + K_0\delta(jt) \tag{3.A.18}$$

where R_o is the residue of $P^{-1}(t)N_o(t)$ in t=0. Note that the constant behavior of the MF inverse transform is related to the delta function.

The inverse transform yields

$$x(w) = a\ sign(jw) + b - \frac{j}{2\pi}\int_{-j\infty}^{j\infty} \left\{ P^{-1}(t)N_o(t) - \frac{R_o}{jt} \right\} e^{-jt(w\pm\varphi)}dt \tag{3.A.19}$$

where the integral is vanishing for $w \to \pm j\infty$ and a and b are related to R_o and K_o, which can be evaluated by imposing

$$\mathrm{x}(j\infty) = \mathrm{a\,sign}(-\infty) + \mathrm{b} = -\mathrm{a} + \mathrm{b} \tag{3.A.20}$$

$$\mathrm{x}(-j\infty) = \mathrm{a\,sign}(\infty) + \mathrm{b} = \mathrm{a} + \mathrm{b} \tag{3.A.21}$$

i.e. $a = [x(-j\infty) - x(j\infty)]/2, b = [\mathrm{x}(-j\infty) + \mathrm{x}(j\infty)]/2$.

It yields

$$x(w) = \frac{x(j\infty) + x(-j\infty)}{2} - \frac{j}{2\pi}\int_{-j\infty}^{j\infty} \{P^{-1}(t)N_o(t)\} e^{-jt(w\pm\varphi)}dt \tag{3.A.22}$$

and taking into account the property (3.A.13), we have

$$x(w) = -\frac{j}{2\pi}\int_{-j\infty}^{j\infty} \{P^{-1}(t)N_o(t)\} e^{-jt(w\pm\varphi)}dt \tag{3.A.23}$$

where the integral is assumed in the sense of Cauchy's principal value.

References

Abrahams, I.D. and Wickham, G.R. (1988). On the scattering of sound by two semi-infinite parallel staggered plates I: Explicit matrix Wiener–Hopf factorization. *Proceedings of the Royal Society of London. Series A*, 420, 131–156.

Abramovitz, M. and Stegun, I. (1972). *Handbook of Mathematical Functions*. Dover Publications, New York, USA.

Achenbach, J.D. (1973). *Wave Propagation in Elastic Solid*. Elsevier, Amsterdam, The Netherlands.

Albani, M. (2007). On Radlow's quarter-plane diffraction solution. *Radio Science*, 42, RS6S11, 1–10.

Antipov, Y.A. and Silvestrov, V.V. (2004a). Second-order functional-difference equations II: Scattering from a right-angled conductive wedge for E-polarization. *Quarterly Journal of Mechanics and Applied Mathematics*, 57(2), 267–313.

Antipov, Y.A. and Silvestrov, V.V. (2004b). Vector functional-difference equation in electromagnetic scattering. *IMA Journal of Applied Mathematics*, 69(1), 27–69.

Antipov, Y.A. and Silvestrov, V.V. (2006). Electromagnetic scattering from an anisotropic half-plane at oblique incidence: The exact solution. *Quarterly Journal of Mechanics and Applied Mathematics*, 59, 211–251.

Auld, B.A. (1973). *Acoustic Fields and Waves in Solids I*. John Wiley and Sons, New York, USA.

Babich, V.M., Lyalinov, M.A. and Grikurov, V.E. (2008). *Diffraction Theory: The Sommerfeld-Malyuzhinets Technique*. Alpha Science, Oxford, UK.

Bernard, J.M.L. (1998). Diffraction at skew incidence by an anisotropic impedance wedge in electromagnetic theory, a new class of canonical cases. *Journal of Physics A: Mathematical and General*, 31(2), 595–613.

Borovikov, V.A. (1998). Diffraction by an impedance wedge with curved sides. *Journal of Communications Technology and Electronics*, 43(12), 1337–1346.

Bresler, A.D. and Marcuvitz, N. (1956). Operator methods in electromagnetic field theory. Report R-495,56, PIB-425, MRI Polytechnic Institute of Brooklyn, pp. 34–36.

Budaev, B. (1995). *Diffraction by Wedges*. Longman Scientific & Technical, Burnt Mill, Harlow, UK.

Daniele, V.G. (1971). Wave propagation in stratified multifluid plasma. *Alta Frequenza*, 40, 904–914.

Daniele, V.G. (2001). New analytical methods for wedge problems. *2001 International Conference on Electromagnetics in Advanced Applications (ICEAA01)*, September 10-14 2001, Torino, Italy, pp. 385–393.

Daniele, V.G. (2002). Two new methods for wedge problems. *Atti della Fondazione Giorgio Ronchi*, LVII(4), 629–650.

Daniele, V.G. (2003a). Rotating waves in the Laplace domain for angular regions. *Electromagnetics*, 23(3), 223–236.

Daniele, V.G. (2003b). The Wiener–Hopf technique for impenetrable wedges having arbitrary aperture angle. *SIAM Journal of Applied Mathematics*, 63, 1442–1460.

Daniele, V.G. (2004). The Wiener–Hopf technique for wedge problems. Dipartimento di Elettronica, Politecnico di Torino, Internal Rep. ELT-2004-2 [Online]. Available: http://personal.det.polito.it/vito.daniele.

Daniele, V.G. and Lombardi, G. (2005). The Wiener–Hopf technique for impenetrable wedge problems. *International Conference on Days on Diffraction 2005*, Saint Petersburg, Russia, pp. 50–61.

Daniele, V. and Lombardi, G. (2006). Wiener–Hopf solution for impenetrable wedges at skew incidence. *IEEE Transactions on Antennas and Propagation*, 54(9), 2472–2485.

Daniele, V.G. and Lombardi, G. (2007). Fredholm factorization of Wiener–Hopf scalar and matrix kernels. *Radio Science*, 42(RS6S01), 1–19.

Daniele, V.G. and Lombardi, G. (2016). Arbitrarily oriented perfectly conducting wedge over a dielectric half-space: Diffraction and total far field. *IEEE Transactions on Antennas and Propagation*, 64(4), 1416–1433.

Daniele, V.G. and Zich, R.S. (2014). *The Wiener–Hopf Method in Electromagnetics*. Scitech Publishing, Edison, NJ, USA.

Daniele, V.G., Lombardi, G. and Zich, R.S. (2017a). The electromagnetic field for a PEC wedge over a grounded dielectric slab 1: Formulation and validation. *Radio Science*, 52, 1–20.

Daniele, V.G., Lombardi, G. and Zich, R.S. (2017b). The electromagnetic field for a PEC wedge over a grounded dielectric slab 2: Diffraction, modal field, surface waves, and leaky waves. *Radio Science*, 52, 1–18.

Daniele, V.G., Lombardi, G. and Zich, R.S. (2017c). Network representations of angular regions for electromagnetic scattering. *PLoS One*, 12(8), 1–53.

Daniele, V.G., Lombardi, G. and Zich, R.S. (2018). The double PEC wedge problem: Diffraction and total far field. *IEEE Transactions on Antennas and Propagation*, 66(12), 6470–6673.

Felsen, L.B. and Marcuvitz, N. (1973). *Radiation and Scattering of Waves*. Prentice-Hall Inc., Englewood Cliffs, NJ, USA.

Friedman, B. (1956). *Principles and Techniques of Applied Mathematics*. Dover Publications New York, USA.

Gakhov, F. (1990). Chapter 9. In *Boundary Value Problems*. Dover Publications, New York, USA.

Gohberg, I. and Krupnik, N. (1992). *One Dimensional Linear Singular Integral Equation I*. Birkhauser Verlag, Basil-Boston-Berlin.

Jones, D.S. (1964). *The Theory of Electromagnetism*. Pergamon Press, Oxford, UK.

Kantorovich, L.V. and Krylov, V.I. (1964). *Approximate Methods of Higher Analysis*. Noordhoff, Groningen, The Netherlands.

Keller, J.B. (1979). Progress and prospects in the theory of linear waves propagation. *SIAM Review*, 21(2), 229–245.

Kong, J.A. (1975). *Theory of Electromagnetic Waves*. Wiley, New York, USA.

Krasnov, M.L., Kiselev, A.K. and Makarenko, G.I. (1983). *Integral Equations*. Mir, Moscow, Russia.

Kraut, E.A. and Lehaman, G.W. (1969). Diffraction of electromagnetic waves by a right-angled dielectric wedge. *Journal of Mathematical Physics*, 10, 1340–1348.

Lyalinov, M.A. and Zhu, N.Y. (2013). *Scattering of Waves by Wedges and Cones with Impedance Boundary Conditions*. Scitech Publishing, IET, p. 215.

MacDonald, H.M. (1915). A class of diffraction problems. *Proceedings of the London Mathematical Society*, 14, 419–427.

Malyuzhinets, G.D. (1958a). Excitation, reflection and emission of surface waves from a wedge with given face impedance. *Soviet Physics Doklady*, 3, 752–755.

Malyuzhinets, G.D. (1958b). Relation between the inversion formula for the Sommerfeld integral and the formulas of Kontorovich–Lebedev. *Soviet Physics Doklady*, 3, 266–268.

Noble, B. (1958). *Methods Based on the Wiener–Hopf Technique for the Solution of Partial Differential Equations*. Pergamon Press, Belfast, Northern Ireland.

Oliner, A.A. (1969). Microwave network methods for guided elastic waves. *IEEE Transactions on Microwave Theory and Techniques*, 17(11), 812–826.

Osipov, A.V. (1993). On the method of Kontorovich–Lebedev integrals in problems of wave diffraction in sectorial media. *Problems of Diffraction and Propagation of Waves*, 25, 173–219.

Peters, A.S. (1950). The effect of a floating mat on water waves. *Commun. Pure Appl. Math.*, 3(3), 319–354.

Peters, A.S. (1952). Water waves over sloping beaches and the solution of a mixed boundary value problem for scalar Helmholtz equation in a sector. *Communications on Pure and Applied Mathematics*, 5, 97–108.

Poincaré, H. (1892). Sur la polarization par diffraction. *Acta Mathematica*, 16, 297–339.

Radlow, J. (1964). Diffraction by a right-angled dielectric wedge. *International Journal of Engineering Science*, 2, 275–290.

Rawlins, A.D. (1977). Diffraction by a dielectric wedge. *Journal of the Institute of Mathematics and its Applications*, 19, 231–279.

Salem, M.A., Kamel, A. and Osipov, A.V. (2006). Electromagnetic fields in the presence of an infinite dielectric wedge. *Proceedings of the Royal Society A: Mathematical, Physical and Engineering Sciences*, 462, 2503–2522.

Senior, T.B.A. (1999). Second order difference equation in diffraction theory. *XXVI URSI GASS*, Toronto, Canada, 64, pp. 13–21.

Senior T.B.A. and Legault, S.R. (1998). Diffraction by an anisotropic impedance half-plane at skew incidence. *Electromagnetics*, 18, 207–225.

Senior, T.B.A. and Volakis, J.L. (1995). *Approximate Boundary Conditions in Electromagnetics*. IEE Electromagnetic Wave Series, London, UK, p. 353.

Sommerfeld, A. (1896). Mathematische theorie der diffraktion. *Mathematische Annalen*, 47, 317–341.

van der Pol, B. and Bremmer, H. (1955). *Operational Calculus*. Cambridge Press, Chelsea, New York, USA.

Vekua, N.P. (1967). *Systems of Singular Integral Equations*. Noordhoff, Groningen, The Netherlands.

Wiener, N. and Hopf, E. (1931). *Uber eine Klasse singularer Integralgleichungen*. S.B. Preuss. Akad. Wiss., pp. 696–706.

Williams, W.E. (1959). Diffraction of an E-polarized plane wave by an imperfectly conducting wedge. *Proceedings of the Royal Society of London. Series A*, 252, 376–393.

Index

G, H, I

K, L, M

N, O, P

R, S, T

U, V, W

Summary of Volume 2

Chapter 5. Fredholm Factorization Solutions of GWHEs for the Electromagnetic Impedance Wedges Surrounded by an Isotropic Medium

Chapter 6. Diffraction by Penetrable Wedges

Other titles from

iSTE

in

Waves

2020

SAKHO Ibrahima
Introduction to Quantum Mechanics 2: Wave-Corpuscle, Quantization & Schrödinger's Equation

2019

BERTRAND Pierre, DEL SARTO Daniele, GHIZZO Alain
The Vlasov Equation 1: History and General Properties

DAHOO Pierre-Richard, LAKHLIFI Azzedine
Infrared Spectroscopy of Triatomics for Space Observation
(Infrared Spectroscopy Set – Volume 2)

RÉVEILLAC Jean-Michel
Electronic Music Machines: The New Musical Instruments

ROMERO-GARCIA Vicente, HLADKY-HENNION Anne-Christine
Fundamentals and Applications of Acoustic Metamaterials: From Seismic to Radio Frequency
(Metamaterials Applied to Waves Set – Volume 1)

SAKHO Ibrahima
Introduction to Quantum Mechanics 1: Thermal Radiation and Experimental Facts Regarding the Quantization of Matter

2018

SAKHO Ibrahima
Screening Constant by Unit Nuclear Charge Method: Description and Application to the Photoionization of Atomic Systems

2017

DAHOO Pierre-Richard, LAKHLIFI Azzedine
Infrared Spectroscopy of Diatomics for Space Observation (Infrared Spectroscopy Set – Volume 1)

PARET Dominique, HUON Jean-Paul
Secure Connected Objects

PARET Dominque, SIBONY Serge
Musical Techniques: Frequencies and Harmony

RÉVEILLAC Jean-Michel
Analog and Digital Sound Processing

STAEBLER Patrick
Human Exposure to Electromagnetic Fields

2016

ANSELMET Fabien, MATTEI Pierre-Olivier
Acoustics, Aeroacoustics and Vibrations

BAUDRAND Henri, TITAOUINE Mohammed, RAVEU Nathalie
The Wave Concept in Electromagnetism and Circuits: Theory and Applications

PARET Dominique
Antennas Designs for NFC Devices

PARET Dominique
Design Constraints for NFC Devices

WIART Joe
Radio-Frequency Human Exposure Assessment

2015

PICART Pascal
New Techniques in Digital Holography

2014

APPRIOU Alain
Uncertainty Theories and Multisensor Data Fusion

JARRY Pierre, BENEAT Jacques N.
RF and Microwave Electromagnetism

LAHEURTE Jean-Marc
UHF RFID Technologies for Identification and Traceability

SAVAUX Vincent, LOUËT Yves
MMSE-based Algorithm for Joint Signal Detection, Channel and Noise Variance Estimation for OFDM Systems

THOMAS Jean-Hugh, YAAKOUBI Nourdin
New Sensors and Processing Chain

TING Michael
Molecular Imaging in Nano MRI

VALIÈRE Jean-Christophe
Acoustic Particle Velocity Measurements using Laser: Principles, Signal Processing and Applications

VANBÉSIEN Olivier, CENTENO Emmanuel
Dispersion Engineering for Integrated Nanophotonics

2013

BENMAMMAR Badr, AMRAOUI Asma
Radio Resource Allocation and Dynamic Spectrum Access

BOURLIER Christophe, PINEL Nicolas, KUBICKÉ Gildas
Method of Moments for 2D Scattering Problems: Basic Concepts and Applications

GOURE Jean-Pierre
Optics in Instruments: Applications in Biology and Medicine

LAZAROV Andon, KOSTADINOV Todor Pavlov
Bistatic SAR/GISAR/FISAR Theory Algorithms and Program Implementation

LHEURETTE Eric
Metamaterials and Wave Control

PINEL Nicolas, BOURLIER Christophe
Electromagnetic Wave Scattering from Random Rough Surfaces: Asymptotic Models

SHINOHARA Naoki
Wireless Power Transfer via Radiowaves

TERRE Michel, PISCHELLA Mylène, VIVIER Emmanuelle
Wireless Telecommunication Systems

2012

LALAUZE René
Chemical Sensors and Biosensors

LE MENN Marc
Instrumentation and Metrology in Oceanography

LI Jun-chang, PICART Pascal
Digital Holography

2011

BECHERRAWY Tamer
Mechanical and Electromagnetic Vibrations and Waves

BESNIER Philippe, DÉMOULIN Bernard
Electromagnetic Reverberation Chambers

GOURE Jean-Pierre
Optics in Instruments

GROUS Ammar
Applied Metrology for Manufacturing Engineering

LE CHEVALIER François, LESSELIER Dominique, STARAJ Robert
Non-standard Antennas

2010

BEGAUD Xavier
Ultra Wide Band Antennas

MARAGE Jean-Paul, MORI Yvon
Sonar and Underwater Acoustics

2009

BOUDRIOUA Azzedine
Photonic Waveguides

BRUNEAU Michel, POTEL Catherine
Materials and Acoustics Handbook

DE FORNEL Frédérique, FAVENNEC Pierre-Noël
Measurements using Optic and RF Waves

FRENCH COLLEGE OF METROLOGY
Transverse Disciplines in Metrology

2008

FILIPPI Paul J.T.
Vibrations and Acoustic Radiation of Thin Structures

LALAUZE René
Physical Chemistry of Solid-Gas Interfaces

2007

KUNDU Tribikram
Advanced Ultrasonic Methods for Material and Structure Inspection

PLACKO Dominique
Fundamentals of Instrumentation and Measurement

RIPKA Pavel, TIPEK Alois
Modern Sensors Handbook

2006

BALAGEAS Daniel *et al.*
Structural Health Monitoring

BOUCHET Olivier *et al.*
Free-Space Optics

BRUNEAU Michel, SCELO Thomas
Fundamentals of Acoustics

FRENCH COLLEGE OF METROLOGY
Metrology in Industry

GUILLAUME Philippe
Music and Acoustics

GUYADER Jean-Louis
Vibration in Continuous Media

Printed and bound by CPI Group (UK) Ltd, Croydon, CR0 4YY
16/09/2026
14963075-0004